Make Your Destructive, Dynamic, and Attribute Measurement System Work for You

Also available from ASQ Quality Press:

The Metrology Handbook
Jay Bucher, editor

Design for Six Sigma as Strategic Experimentation: Planning, Designing, and Building World-Class Products and Services
H. E. Cook

Concepts for R&R Studies, Second Edition
Larry B. Barrentine

Failure Mode and Effect Analysis: FMEA From Theory to Execution, Second Edition
D. H. Stamatis

The Uncertainty of Measurements: Physical and Chemical Metrology: Impact and Analysis
S. K. Kimothi

Integrating Inspection Management into Your Quality Improvement System
William D. Mawby

The Certified Manager of Quality/Organizational Excellence Handbook, Third Edition
Russell T. Westcott, editor

Making Change Work: Practical Tools for Overcoming Human Resistance to Change
Brien Palmer

Business Performance through Lean Six Sigma: Linking the Knowledge Worker, the Twelve Pillars, and Baldrige
James T. Schutta

Transactional Six Sigma for Green Belts: Maximizing Service and Manufacturing Processes
Samuel E. Windsor

Enterprise Process Mapping: Integrating Systems for Compliance and Business Excellence
Charles G. Cobb

To request a complimentary catalog of ASQ Quality Press publications, call 800-248-1946, or visit our Web site at http://qualitypress.asq.org.

Make Your Destructive, Dynamic, and Attribute Measurement System Work for You

William D. Mawby

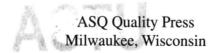
ASQ Quality Press
Milwaukee, Wisconsin

American Society for Quality, Quality Press, Milwaukee 53203
© 2006 by American Society for Quality
All rights reserved. Published 2006
Printed in the United States of America
12 11 10 09 08 07 06   5 4 3 2 1

Library of Congress Cataloging-in-Publication Data

Mawby, William D., 1952–
   Make your destructive, dynamic, and attribute measurement system work for
   you / William D. Mawby.
      p. cm.
   Includes bibliographical references and index.
   ISBN-13: 978-0-87389-691-7 (alk. paper)
   1. Process control—Statistical methods. 2.   Mensuration.   I. Title.

   TS156.8.M39   2006
   658.5′62—dc22

                                                        2006007817

Publisher: William A. Tony
Acquisitions Editor: Annemieke Hytinen
Project Editor: Paul O'Mara
Production Administrator: Randall Benson

ASQ Mission: The American Society for Quality advances individual, organiza-
tional, and community excellence worldwide through learning, quality improvement, and knowledge exchange.

Attention Bookstores, Wholesalers, Schools, and Corporations: ASQ Quality
Press books, videotapes, audiotapes, and software are available at quantity
discounts with bulk purchases for business, educational, or instructional use. For
information, please contact ASQ Quality Press at 800-248-1946, or write to ASQ
Quality Press, P.O. Box 3005, Milwaukee, WI 53201-3005.

To place orders or to request a free copy of the ASQ Quality Press Publications
Catalog, including ASQ membership information, call 800-248-1946. Visit our
Web site at www.asq.org or http://qualitypress.asq.org.

Quality Press
600 N. Plankinton Avenue
Milwaukee, Wisconsin 53203
Call toll free 800-248-1946
Fax 414-272-1734
www.asq.org
http://qualitypress.asq.org
http://standardsgroup.asq.org
E-mail: authors@asq.org

∞ Printed on acid-free paper

I dedicate this work to my daughter Briana and wife LuAnne, who make it all possible for me every day and in every way.

I also dedicate this work to all the practitioners who have struggled with applying measurement systems in realistic situations. I sincerely hope that the information in this book will enable you to achieve a personal level of success in making real improvements to your company's quality performance.

Contents

Figures and Tables

Preface

The motivation for this book comes from my own extensive experiences with trying to apply the standard measurement systems analysis (MSA) methods to real-world problems that arise in manufacturing and industry. The methods work well for simple systems that are not dynamic and that have only two significant sources of error, but they leave a lot to be desired in other circumstances. This shortfall is especially clear in the case of attribute MSA, in which the standard method is far from adequate.

This book provides clear procedures for situations in which the part values change or are destroyed. It also provides procedures that work when the measurements are dynamic and cannot be separated from the process. It extends the simple methods to cases in which the measurement systems have several sources of uncertainty. And it completely overhauls the attribute methodology and enables it for many difficult but practical applications. Each extension of the method is detailed in a chapter complete with realistic examples and a summary called "Take-Home Pay," which is provided to clue the reader into the key points that are critical for the attempt to enable bottom-line success. Readers who use these methods will find that they can quickly make significant improvement in their destructive, dynamic, and attribute measurement systems with less effort.

Chapter 1 outlines the need for good measurements and shows, with examples, how errors can adversely affect several common decision processes. Measurement errors can affect product classifications by causing false leaks and false alarms. They can also affect the control of processes both through statistical control and engineering control systems by inflating control limits. Measurement errors can impact process improvement by adding uncertainty to risks and rewards. And they can impact the learning process by increasing the number of false steps that must be endured when traveling from one phase to another.

Chapter 2: There are various approaches to the study of measurement systems, including the Automobile Industry Action Group's manual *Measurement Systems Analysis,* third edition. Because the AIAG guide is the standard for so many industries in North America, it is used as the base analysis for further discussion in this book. In this guide there are two approaches given to the estimation of device and operator effects—that is, repeatability and reproducibility. One method uses a complicated calculation involving ranges and averages while a more straightforward analysis of variance (ANOVA) is also given. This chapter details the application of the ANOVA approach to variables MSA that serves as the method of choice for further examination and evaluation.

Chapter 3: Building on the ANOVA approach to estimating the measurement error components, a formula is given to compute the standard deviation of the estimates of the different measurement errors including parts, repeatability, reproducibility, and RandR. In this way it is shown that both the sample size and the allocation of samples to operators drastically affect the variability. Several scenarios are then derived both theoretically and through simulation to show the effects of increasing parts, operator, or repeats. Additionally, a discussion of using this information to generate the best designs and the best cost designs for variables MSAs is given.

The standard MSA examines only the effects of operator and repeats by themselves. Chapter 4 details through simulation the meaning and requirement for adding parts by operator interaction. It also offers motivation for extending the MSA to other sources of dispersion, as recommended in a measurement uncertainty analysis. A discussion is given of measurement uncertainty analysis at this point. The advantage of using these other factors is shown for understanding the system better and for achieving better estimates of all the measurement error components. Other interactions between these additional factors are then exampled and examined for their impact.

Chapter 5: The standard variable MSA study applies satisfactorily to systems in which the part value does not change from repeat to repeat. But in many systems the true value of each part cannot be held constant between repeats. Deformative systems are systems in which repeats are possible, but the true value changes. But this change, this deformation, is not enough to destroy the measured product completely and therefore is not destructive. The inflation and misattribution of measurement error by product changes is demonstrated through a series of simulations. Several methods for attacking this problem are discussed, including the naive approach, a change-of-design approach, and a covariance approach. It is shown that this covariance approach, whether applied in a two-step or a one-step method, can perform well and offers great potential for other difficult MSA situations as well.

Chapter 6: Another problem for the standard MSA study occurs when the measurement actually destroys the part and makes any repeat measurement impossible. Again, it is shown through simulation that the effect of this problem is often to inflate or contaminate estimates of measurement error with product variation. Several solutions to this problem are discussed through the use of detailed examples including the naive approach, a design approach, and a covariate approach. The covariate approach is extended as well to more complex situations and scenarios in which the pattern must be part of statistical discovery.

Chapter 7: In some measurement situations, the parts are not destroyed or deformed during the measurement process; nevertheless, they cannot be repeated because the system is dynamic. These kinds of dynamic measurements occur with in-line scales, many chemical tests, and other dynamic situations. The fundamental problem is that the measurement systems get only one shot at a particular product, and again, no true repeat is possible. The impact of this problem on measurement error contamination is illustrated. Several possible solutions for this situation are described, including standards, splitting, matching, and covariance analysis.

With the arguments given in the earlier chapters, it is clear that the design of the measurement study is critical for success. Chapter 8 summarizes the lessons learned for design, including the increase of operators and the choice of total sample size. Other designs are exampled, including MSAs based on factorial designs, nested design, and covariance designs. There is also a discussion of the choice of part values to decrease sample size. Also studied are the effects that using a different design has on meeting the standard MSA targets.

Chapter 9: Although variables measurements appear to be the most common types of systems subjected to MSA study, it can be argued that, in practice, it is attribute systems that are more common and potentially more dangerous. The standard attribute MSA prescribed by the AIAG manual is used to orient and drive discussion of the issues. This method consists of special selection of 50 parts on which three operators perform three repeats. The results of this study are then analyzed by counting the frequency of important events such as false alarms, leaks, and mixed readings. The targets for acceptable performance are described, and it is shown through simulated example that sample size, sample allocations, and part selection have dramatic effects on the results.

In Chapter 10, a different approach to attribute MSA is given. This method fits a model with parameters representing various important probabilities such as the probability of an acceptable part, the probability of a leak, and the probability of the correct identification of an acceptable part. Through examples, it is shown how this method can be used to study repeatability and operator effects for the cases in which standards may or may not be available. Several applications of this approach are then detailed, including the study of additional sources of error and the grading of parts through this method.

Chapter 11: The use of logistic regression allows one to extend the attribute measurement analysis to alternate and additional sources of uncertainty. It also allows one to perform the attribute analysis on deformative, destructive, and dynamic situations. Examples are given of the various extensions.

Measurement systems are often so ubiquitous that they get ignored. As is shown, the impact of poor measurement systems can be rather large. This can be the case when there are a few expensive decisions made or when there are many replicated small decisions. Chapter 12 demonstrates a method for converting measurement system error plus process capability into an associated risk analysis. The gage performance curve and how this curve forms the basis for the risk analysis is discussed. Ideas and suggestions for starting an evaluation and monitoring system for MSAs are described. It is possible that such tracking is essential for achieving good measurement quality assurance in a company.

As difficult as it often is to get an initial MSA study performed, it is even more difficult to get follow-ups done. In Chapter 13, a discussion is given of what measurement systems analyses need to be updated. Also, several methods are given that can help organize this updating process. These methods include periodic updates, triggers based on auditing or reduced testing, reliability-based approaches, control chart approaches, and cost-based approaches. Indicators of measurement systems performance are proposed that try to track results through time and make proper summaries of impacts through meta-analytic or other approaches.

The different features of the book are reviewed in Chapter 14. First, the impacts of measurement error on decisions are discussed. Then, an accounting of the standard MSA is taken, which leads to modifications and extensions of the method to additional sources of measurement uncertainty. The issues of the design of MSA studies are then discussed. Next, a very important extension of the MSA method to deformative, destructive, and dynamic systems is made. Attribute MSA is discussed and extended with two different approaches. The logistic-regression-based approach allows an enormous extension of the analysis to deformative, destructive, and dynamic systems as well. Then, the evaluation and maintenance of measurement systems is described. Once the summary is complete, a short discussion is given that attempts a prediction of future MSA needs.

Some of the methods that are required to overcome and extend the standard MSA to destructive, dynamic, and attribute systems require a reasonable level of quality engineering knowledge. Although it is certainly possible (and indeed an objective of the book) that any reader can understand the approaches described, it will certainly take some work to apply the methods to new situations. Some of the prerequisites for the reader who is aiming to take full advantage of this material in the book include familiarity with the standard MSA, familiarity with ANOVA methods, familiarity with covariate analysis, and familiarity with logistic regression. Some other techniques (like Fourier analysis, reliability analysis, and risk computations) are used but are not critical to the main themes of the book.

1

The Need for Measurement Systems Analysis

THE CRITICAL NATURE OF SENSES

Measurements provide the only way to know the world. It is only through sensors, and ultimately senses, that one can interact with the physical world. This is true regardless of whether the measurement is a viewing of a game of baseball or the monitoring of a manufacturing process. If the measurements are contaminated with error, they could result in wrong decisions. Assuming that it is impossible to completely eliminate all of the measurement error, the decision maker has only two broad alternatives to consider, as shown in Figure 1.1. One alternative is to ignore the fact that there is any measurement error at all and act as if the data are completely and utterly

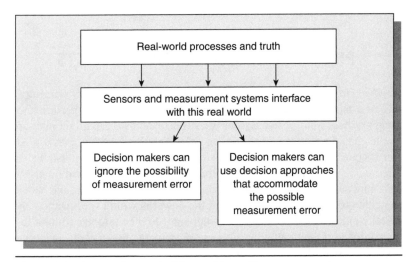

Figure 1.1 Approaches to decision making with measurement error.

1

trustworthy. This is clearly a foolhardy approach if the measurement error causes problems. Errors made in cancer screening or in safety checks on an amusement park ride could have catastrophic results under this approach. The other approach is to acknowledge the measurement errors and adjust for them in the decision-making process. When the error can be considered random, the field of statistics can provide good, time-tested methods for doing this adjustment properly (Dixon, 1983). Statistical technology consists of methods and tools whose sole purpose is to enable one to achieve good decisions in spite of the ever-present measurement variations.

Because measurement systems play such an intimate role in the interactions of human beings with their environment, it is small wonder that there are countless ways in which measurement errors impact decisions. Even restricting the investigation to only manufacturing and business processes will still leave many challenges. However, lessons that one learns in this narrower application should be easy to back-propagate to other areas. In manufacturing and business processes, the measurements are typically of three types: through *human senses alone,* through *instrumentation alone,* and through *instrumentation-aided human senses.*

As little as 20 years ago, the first of these options (the one depending solely on human senses) was far and away the most common. But it must be said that today, due to the explosion in the use of computers, electronics, and software, the second and third options are catching up and in some cases surpassing the human-based measurement systems in their frequency of application. These measurement systems, regardless of their type, are used for at least four purposes: (1) *product classification,* (2) *process adjustment and control,* (3) *data analysis,* and (4) *reasoning and learning.* Each of these applications has its own peculiarities of impact, which will be examined in detail.

PRODUCT CLASSIFICATION EFFECTS

Product classification is the activity of measuring one or more characteristics of a particular product and then assigning that item to a performance category based on those characteristics. For example, one might measure the length of a wooden dowel rod and compare this length to preset product specifications dictated by the customer. This process is illustrated in Figure 1.2. If one assumed that the true product length is 9.9 cm (unknowable in a real situation) and the product specification is 10 cm, one would conclude (correctly in this case) that the dowel is within specification and usually one would feel justified in shipping it to the customer without further inspection and certainly without any rework. But if there is a measurement error of +0.2 cm that occurs when one measures this particular part with a digital micrometer, the measured value of the part could read

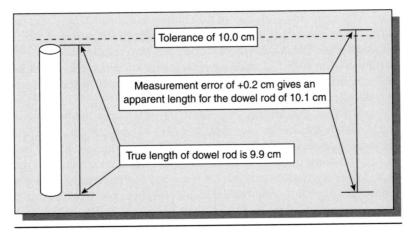

Figure 1.2   A dowel rod is measured as too long.

(erroneously) as 10.1 cm. This measured value is the only knowledge one has on which to act, because one cannot know the true length of the dowel without measuring it. The direct impact of this measurement error is to classify the dowel above the specification limit. Depending on the type of shop and the type of product, the result of this misclassification can be one of many actions. If feasible, it might be possible to remeasure the same part to get a second opinion. Sometimes this second measurement is assumed to replace the first, but it is used solely to direct the product disposition from this point onward. At other times the two measurements are combined into one decision by averaging or voting approaches. Many times this redo of a measurement is allowed to occur a third or fourth time or even more if each added measurement continues to fall outside the tolerance span. Very few companies have policies for remeasurement of parts that fall within the specifications. Intuitively, there should be some best or optimal number of remeasurements that are allowable, because each one requires more time and effort and hence more money to support it (Ding, 1998).

This kind of in-line remeasurement has a cost, but it is low in most cases compared with that caused by other possible measurement-error-induced actions. Because operational speeds and cycle times often depend on the process running without interruption, products measured as nonconforming cannot be easily measured in situ and must instead be dropped into an alternate repair or reinspection pathway. Once a part enters this alternate path, the part may be reinspected more carefully, sent for repair, or even scrapped. The cost induced by this extra activity is not negligible and in the case of a false scrap can be pretty penalizing. This cost, whatever it might be, can be directly blamed on the presence of measurement error. Keep in mind that a false scrap wastes the cost of materials and any labor expended in the manufacture of the particular item up to the point in the process at which it was detected. The probability of making this kind of false scrap is called a *type I error* in statistical studies (Lindlay, 1985). It can also be called the producer's risk or the false negative rate depending on the particular areas of application. It is shown in Figure 1.3.

Still within the category of product classification is another kind of costly mistake that can be caused by measurement error. Assume the part, the wooden dowel rod, is truly 10.1 cm. According to the description given earlier, this dowel should rightly be classified as nonconforming, and one should stop any further work on this piece and begin corrective action or rework. However, there could be a measurement error of –0.2 cm that causes the measured value of this particular dowel rod to read as 9.9 cm. This measurement value would make it look as if the dowel rod sits just within the specification. See Figure 1.4 for a depiction of this process. Again, because the true value of the rod's length is not known, it is only this measured value that one has to lead further actions. In this case the measurement error would lead one erroneously to treat the piece as conforming and to ship it out to the customer without further concern. It is certainly possible to protect against this kind of error with repeated measurements of the part in a quality chain approach, but the error will still lead to additional costs. If the out-of-specification part gets shipped to the customer and results in poor performance, the costs could be very high indeed.

This kind of error, leading to leaked nonconforming parts or delaying their discovery, is called *type II* in statistical science (Lindlay, 1985). It is also called consumer's risk or false positive rate in other fields. This type of error can be caused by measurement error, as shown in Figure 1.5.

Both kinds of product classification errors have their biggest effects on products whose true values are close to the specification limits. Products with true values far away from the specification limits will be misclassified

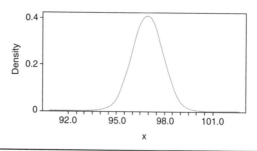

Figure 1.3 Type I or false scrap error versus the true product value.

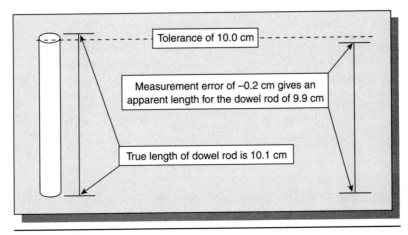

Figure 1.4   A dowel rod is measured as too short.

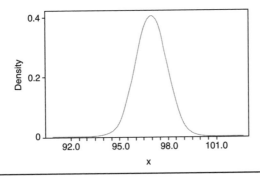

Figure 1.5   Type II or false accept error versus the true product value.

only when truly large, and therefore rare, measurement errors occur. In general, the probability of a misclassification goes down as the true value separates itself from the danger zone clustered around the tolerances. Because the number of parts produced in this danger zone is characterized by the production process capability, the impact of the measurement error is worsened by low capabilities. One can combine the process capability with the measurement capability to produce effective predictions of misclassification errors, as illustrated in Figure 1.6. Programs like Six Sigma aim to decrease the process variability so much that there is very little chance that products are created that fall inside this misclassification danger zone (George, 2002).

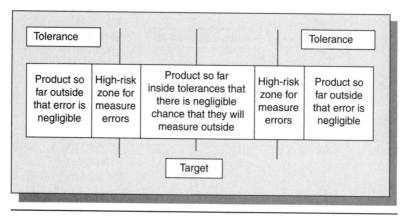

Figure 1.6 The danger zones for increased decision errors.

PROCESS ADJUSTMENT

Measurement error can also impact other kinds of decisions, including those affecting process adjustment policies (Box, 1997). As an instructive example, consider a process that is perfect. That is, every single part produced from this process is precisely on target and there is no deviation from this ideal. This is a wonderful (but artificial) situation, for it means that every single part that this process produces will be just what the customer ordered and what the designers designed. Although this sounds great, there is a problem. Specifically, one cannot know that this process is perfect without measuring it. If the measurement system is not absolutely without error, the perfectly produced dowel rods will not measure as perfect. The rods will measure differently from the target and from each other. The observer of these measured values will not be able to distinguish whether the dispersion is stemming from the process, the measurement process, or both.

This combination of perfect process and imperfect measurement system can still lead to trouble and increased costs if improper interpretation of the error is made. Imagine that this perfect process is activated and the first part is produced. Being perfect, the dowel rod's length is right on target at 100 cm. A measurement is made, with the resulting value of 100.3 cm. If one knew that this additional 0.3 cm was due to measurement error, then one would do well to ignore it and let the process run without any attempt to improve it. But if one did not know that this deviation of 0.3 cm was due to measurement error and instead interpreted it as a process problem, then one might do something foolish. One could try to adjust the process downward by 0.3 cm to compensate for the apparent process overshoot. With the now-adjusted process, the true value of the second part that is produced will be 100 cm − 0.3 cm = 99.7 cm for the assumed perfect process. Notice that the misinterpretation of the sources of the deviation from target has now induced a real process offset, and the process is no longer perfect. This scenario is illustrated in Figure 1.7.

One can continue in this way for the third and fourth parts and so on. Each time, the measurement error will be forced into the process variation. If the measurement error has no bias in it—that is, if it averages to zero over the long term—then the process should also average to the target. But the process will no longer produce every dowel rod consistently at this target. Instead, the process will move in exact proportion to the variation from the measurement error. This is likely to be a series of random jumps of size mostly near zero but with occasional extremes. If the process has some dispersion characterized by standard deviation p and the measurement system is characterized by standard deviation m, the new process standard deviation will equal the square root of $p^2 + m^2$. Notice that this is real process dispersion in the sense that the true lengths are moving around by this amount and not just the measured value of them. There will still be an additional layer of measurement noise on top of this process movement. Figure 1.8 illustrates a hypothetical run of 40 consecutive parts produced according to this scenario. In the figure, T = true length, W = measured length, and R = adjustment value.

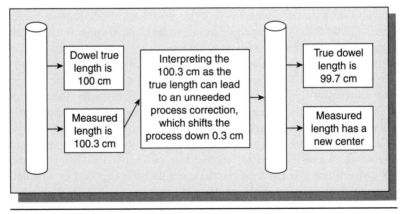

Figure 1.7 Incorrect process compensation due to measurement error.

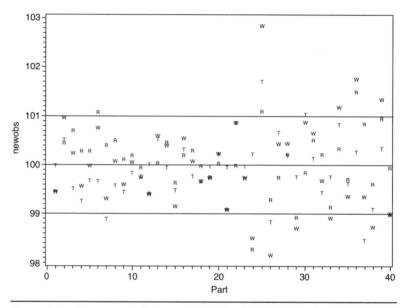

Figure 1.8   A sequence of parts produced with erroneous adjustment.

PROCESS CONTROL EFFECTS

Process control is similar to process adjustment in that some corrective or investigative action is activated whenever a particular signal occurs (Runger, 1991). However, process adjustment is usually nearly instantaneous and smaller in impact, whereas process control usually requires a longer time lag and results in larger structural changes in the process. They are also similar in that measurement error can adversely affect the control of a process as well. The construction of a standard Shewhart process control chart begins with the collection of adequate data to represent the process variation under a statistically stable, or in-control, regimen. The average of this data is used to establish the centerline of the chart, and the control limits are set at plus and minus three standard deviations around the mean. If one wished to plot the average of subgroups of size n, then the limits are set at plus or minus three standard deviations divided by the square root of n. Once it is constructed, the control chart is applied to new data through the plotting of each subgroup average on the chart. If the plotted point is found within the control limits, the process is considered to be unchanged and needs no adjustment. If the point falls beyond either control limit, the process is considered changed or offset from the target, and some action is appropriate to correct this situation. This follow-up action can be a simple remeasurement or the more drastic halt of the process with follow-up investigative work.

The data upon which the control chart is computed are invariably tainted with measurement error. In this way, the standard deviations and the averages are also tainted. Recall the perfect process wherein every product is manufactured right on target at 100 cm with no process variation. A control chart based on the true values of this perfect process would be very strange indeed: All points would fall identically on the target value of 100 cm. It would be impossible to construct proper control limits because the standard deviation (without measurement or process variation) would be zero and thus the control limits would both coincide with the target line.

In this imaginary perfect process it would still be necessary to measure the dowel rods to provide the data necessary to construct and utilize the control chart. In the case where there is a real measurement system in use, the raw data would be the result of the combination of this monotonously good process with the imperfect fluctuations of the measurement system. Again, the average of the long run of these measurements should show an average of 100 cm, but it would have a standard deviation equal to that of the measurement system itself. The control chart centerline will be appropriately placed, and the control limits will capture the variation induced by the measurement system only. Good decisions can be made based on the control charts despite the measurement error.

If the process is not perfect, it will have its own process variability in addition to that coming from the measurement. In this scenario the centerline is still properly placed, but the control limits are spread even wider. A point outside the control limits would be interpreted as a process offset too large to be explained by normal process variation and measurement variation together. Clearly this would be wider than would be necessary if there were no measurement error to reckon. Thus such a measurement-error-infected control chart could lead to a looser control situation in which true process offsets could be allowed before corrective action is taken. That is, the chart has become desensitized by the measurement error. Figure 1.9 illustrates a hypothetical sequence of 18 individual dowel rod lengths on a control chart and compares the control limits that would be set with and without measurement error.

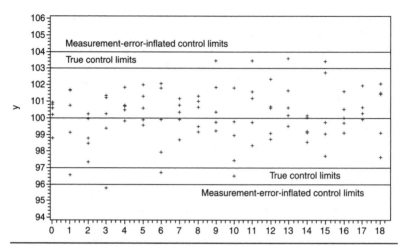

Figure 1.9   Hypothetical control chart on a sequence of lengths.

ANALYSIS EFFECTS

The measurement-error-induced effects on product classification and process adjustment are probably the most obvious effects that can be produced, but there are additional impacts that can be even more costly depending on the circumstances. One of these further impacts is in the area of analysis of the data. Analysis operates on data that exist in historical collections and are often collected from special studies or tests (Harrington, 1991). *Analysis* is the attempt to understand the potentially complex mechanisms of the process in a useful way so that improvements can be effected or new designs created. Typically, studies are limited in scope and infrequent because they tend to have a higher cost-per-item basis than the observational data that are often collected more or less automatically. Consequently, decisions made on the basis of this study data are made on a smaller number of data points than utilized in process control. A few measurement errors affect a relatively higher proportion of this data and can have an inordinate impact on the results of the analysis. Study data are usually so precious that very detailed statistical analysis is the only way in which to extract all the valuable information.

As an example of one possible adverse effect of measurement error on analysis, consider a regression analysis (Ryan, 1989). Two sets of data are taken to test a purported linear relationship. Let's assume that this relationship is perfect; that is, every ordered pair of true values for x and y would fall exactly on a line. There would be little doubt about the relationship between x and y in this case. But if there were appreciable measurement error in the y direction, the measured point pair will not fall on the line but could form a scattered cloud around it. If this variability is large enough it might visually obscure the linear relationship. It might also cause the statistical test to register a low confidence for the linear relationship, and it would keep the R-squared and other gross measures of fit low as well. If the analyses did not show any clear relationships, the investigator would be more likely to abandon this area of inquiry and spend his or her scarce resources on some other approach. In this way a very valuable relationship might be lost. Figure 1.10 shows a hypothetical example of this with the perfect line and the error-induced line.

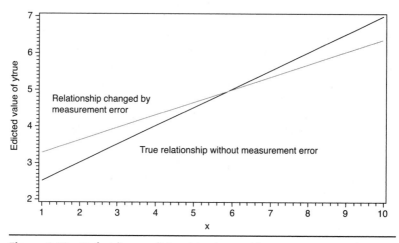

Figure 1.10   Perfect linear relationship obscured by measurement noise.

LEARNING EFFECTS

The full range of effects that measurement error and decision errors can have on the reasoning and learning processes is quite an extensive subject, but only a narrow example of the impact will be treated here. Consider learning as a process whereby one learns the route from the fact at point A to the fact at point B. The rate by which one is able to traverse this route depends on how many mistakes are made along the way. In reality, if this process takes too long or requires too many steps, one is likely to abandon the route entirely, but this aspect of the learning effect will not be considered further here. It is only the effect of measurement error on the number of steps needed to cover the path that will be examined.

Consider a simple example in which there are points on a number line at the integers. The investigator starts at point 0 and then moves up by one number or down by one number at each discrete time. A journey of learning can then be modeled as a sequence of back-and-forth steps from point 0 to point 10 on this grid. In the extreme case of a perfect learning process, the probability of a right (correct) movement is 1, so it should take exactly 10 steps to get to point 10. This ideal shortest path should occur consistently again and again. But if measurement dispersion or error complicates this so that the probability of a right (correct) movement is only 80 percent, the trip will take longer on average and will have a lot more variation. To move to the right consecutively 10 times would occur with the probability $0.80^{10} = 0.107$, so this is not a typical result at all. The average trip length will likely be $10/.80 = 12.5$, but there are some paths, although less probable, that can take an inordinate amount of steps.

Intuitively, the more steps involved in this process, the more likely the chance that the learning process will be slowed. In a more concrete setting, this set of learning steps could be required in order to create a successful process improvement. If the learning is delayed or even abandoned, the improvement will be delayed, and any savings that would have been stimulated will be held in abeyance for that same lag time. If these learning steps are difficult or infrequent, then it is easy to see how erroneous steps may cause entire pathways to be abandoned. At the very least, the measurement error will make the learning process more expensive than it could be otherwise. Figure 1.11 illustrates such a model of a learning pathway and an example pathway through it.

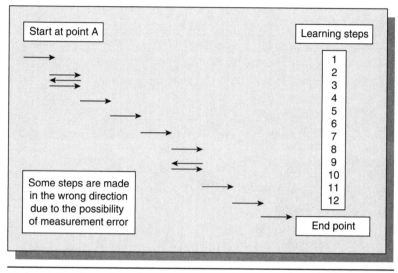

Figure 1.11 The effect of measurement error on the learning pathway.

Chapter 1 Take-Home Pay

1. Measurement errors can adversely affect decisions.

2. Measurement errors can result in product misclassifications.

3. Measurement errors can result in lack of process control.

4. Measurement errors can disrupt statistical model building.

5. Measurement errors can slow learning.

2

Review of the Standard Variables Measurement Systems Analysis

MEASUREMENT UNCERTAINTY

Because the impact of a measurement is so potentially harmful, it is important to understand how one can characterize and comprehend the broader, more inclusive aspects of measurement uncertainty. *Measurement uncertainty* is an attempt to systematically and exhaustively evaluate the potential impact of all sources of measurement error on the variation in a measurement result (AIAG, 2002). The goal is to be able to supply the customer with a useful guarantee on the uncertainty of a particular measurement. Typically, it is recommended that this uncertainty be expressed as an interval with upper and lower limits that fit around the measured value. To produce this interval, it is necessary to provide an estimate of the variability of each component of measurement error that could contribute to the overall measurement uncertainty. Estimates based on expert opinion or data analysis are both legitimate for an uncertainty analysis. Once the uncertainty components have been estimated, they must be combined in a logical manner. Although it is optimal to create a solid model of the ways the uncertainties combine, it is quite common to make an approximation using just a sum-of-squares approach. Finally, this overall or system uncertainty is bracketed in an interval of two or three multiples to account for some of the imprecision in its own estimation. Although the number that is quoted as the system uncertainty can be very important, the procedure also has great value in that it forces the people who are expert with the measurement system to think deeply and clearly about the process. If these same experts continue to stay engaged in this process, the measurement uncertainty will almost continually get better as well. See Figure 2.1 for an illustration of the uncertainty concept.

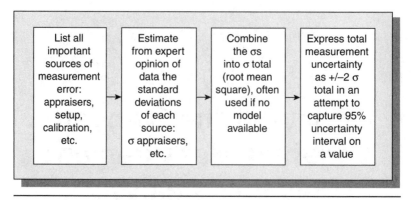

List all important sources of measurement error: appraisers, setup, calibration, etc.	Estimate from expert opinion of data the standard deviations of each source: σ appraisers, etc.	Combine the σs into σ total (root mean square), often used if no model available	Express total measurement uncertainty as +/–2 σ total in an attempt to capture 95% uncertainty interval on a value

Figure 2.1 The measurement uncertainty approach.

MEASUREMENT ANALYSIS

Measurement systems analysis (MSA) is a subset of full measurement uncertainty (Burdick, 2003). Instead of attempting to label and quantify all potential sources of error, as does uncertainty analysis, an MSA targets two components that are consistently expected to have a large effect. These two components are termed *repeatability* and *reproducibility*. *Repeatability* is the effect of those errors that occur under homogeneous conditions and is dominated by instrument-induced error. *Reproducibility* measures the effect of errors associated with operator or appraiser differences. There may be certain applications that are indeed dominated by just these two sources of measurement error, but it is more likely that there are many measurement systems that will have other sources in addition to these two. For example, setup is a common measurement factor that is ignored in this approach. Nevertheless, the MSA methods described in good detail in the Automobile Industry Action Group's manual (AIAG, 2002) are the de facto standard for these kinds of analyses in a great many industries today. This concept is shown in Figure 2.2.

Statistically, the approach is based on the need to estimate components of variance for repeatability and reproducibility. The fundamental model is that each individual measurement is composed of independent terms from appraisers, from parts, and from repetitions. The part effect is assumed to be constant whenever the same part is used, and the same is assumed true for the appraiser effects as well. Only the repetition effect is assumed to change on every unique measurement. Each effect is assumed to be an independent draw from a normal distribution with a mean of zero and a standard deviation char-

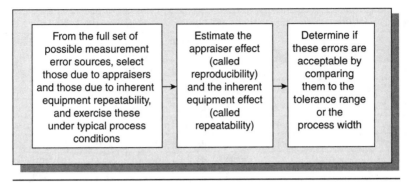

Figure 2.2 The MSA approach.

acteristic of the particular error source. It is the purpose of an MSA to estimate these standard deviations as closely as possible from the data at hand.

Measurement System Design

The design that is recommended for the assessment of these two components of measurement variation is a full factorial based on the three factors of appraisers, parts, and repetitions (Montgomery, 2000). Each appraiser is to measure each part three times in repetition. A typical MSA study might employ 10 different parts measured three times by three appraisers for a total of 90 individual measurements. The AIAG manual provides two methods for estimating the parameters of the model from this full factorial experiment. A *full factorial experiment* is one in which all combinations of levels from all factors are found at least once in the testing. One method is a range-based process that is easier to compute manually but is very restrictive in terms of modifications and extensions. The other method is based on expected mean squares from an analysis of variance (ANOVA) and employs method-of-moments estimation, which will be described in the next chapter and will be used to motivate many explanations and extensions of the method later (Searle, 1971). Expected mean squares are the theoretical counterparts of each line in the ANOVA table and help one disentangle the pure components of variance that are the goal of the analysis.

The data contained in Table 2.1 are that of the standard MSA study with 10 parts, 3 appraisers, and 3 repetitions. Table 2.2 details the results of such an analysis as generated by the ANOVA method, and this will be described in detail in the next chapter.

Table 2.1   Data details for the standard MSA study.

Part	Operator A Repeat 1	Repeat 2	Repeat 3	Operator B Repeat 1	Repeat 2	Repeat 3	Operator C Repeat 1	Repeat 2	Repeat 3
1	103.715	104.230	103.939	102.150	102.137	102.124	102.342	102.695	102.363
2	102.629	102.743	102.598	100.895	101.257	101.298	101.212	101.099	101.350
3	98.103	98.077	97.263	96.536	96.316	95.998	96.518	96.874	96.616
4	95.137	95.281	95.215	93.945	93.641	94.058	94.017	93.929	94.082
5	101.301	101.340	101.318	99.359	99.306	99.110	99.239	99.729	99.629
6	100.560	100.399	100.983	99.400	98.959	99.412	99.339	99.985	99.375
7	98.000	97.720	97.881	96.228	96.219	96.204	96.767	96.622	96.714
8	100.305	100.454	100.414	98.555	98.479	97.985	98.781	98.717	98.603
9	103.235	102.793	102.989	100.968	101.634	101.368	101.888	101.846	101.635
10	101.937	102.142	102.135	100.214	100.668	100.785	100.745	100.815	100.210

Table 2.2   Results of the ANOVA of the standard MSA study.

Error Source	Variance Estimate	Standard Deviation	Percent Tolerance
Parts	7.22	2.69	—
Appraiser	0.78	0.88	26.4
Repeatability	0.05	0.22	6.6
RandR	0.83	0.91	27.3

The Analysis of the Standard MSA

In Table 2.2 there are several quantities that should be of interest to anyone concerned with the quality of this measurement system. First, there is the estimate of the standard deviation of repeatability. This number, 0.22, represents the standard deviation of a normal distribution of possible measurement errors that can be introduced whenever a measurement is made. The act of measurement is seen as a random draw from this normal distribution of errors. In particular, the draws are uncorrelated and have no patterns. In addition, it is assumed that this variance does not change from measurement to measurement. The random draw is from the same pool of numbers, and this pool does not shrink or expand its range of values. Perhaps an even stronger assumption is that these measurement errors follow a normal or bell-shaped curve. Theoretically, this means that any error, even a very large positive or negative one, is possible. And, this assumption means that small errors near zero are more likely than larger ones in a symmetric way.

These assumptions on the properties of measurement error are important because they will determine the targets for acceptability of the measurement system, and this naturally affects real decisions with hard monetary consequences. For example, an inadequate system can lead to a need to acquire or develop a new system. Such an activity can be quite costly and time consuming. On the other hand, if the device is inadequate but remains in service, then each and every measurement can have an unacceptable impact on the decisions made on these devices. These errors can lead to false alarms, leaks of nonconforming products, and many other types of mistakes. Often the costs of these errors might be relatively small for each individual usage, but they can really add up when all the small impacts are accumulated.

Another number of interest in Table 2.2 is the standard deviation of the appraiser effect, or the reproducibility as it is known. This value of 0.88 is larger than that of repeatability but is interpreted in a similar fashion. Namely, it is imagined that there is another draw of a random error from a normal distribution with this standard deviation with mean zero whenever the appraiser changes. Notice that the repeatability effect will, in general, provide a new draw every single measurement, whereas the reproducibility error changes only when a new appraiser is involved. A study with a single operator would have only one error or offset due to appraiser effect for the entire study but would have a new and different repeatability error drawn for each measurement that is made.

On any particular measurement it is assumed that there is random error stemming from the appraiser and a different random error stemming from the repeatability effect. These two errors are assumed to be simply added together to arrive at a final impact on the measurement. It is possible, for example, for the two errors to totally compensate for one another and leave a practically error-free measurement. But it is also possible for these errors to push in the same direction and make the measurement even more incorrect than either error separately would be expected to do. If one imagines this process of picking two errors and adding them together, it is easy to see that the distribution of potential errors will get larger with more variation. It is not easy to see it, but the shape of this combined distribution of errors will still be bell-shaped and will have a resulting standard deviation that is exactly equal to the square root (repeatability standard deviation squared + reproducibility standard deviation squared). This is the idea behind the computation of the RandR, or repeatability and reproducibility, standard deviation.

Measurement Performance Ratios

Because the measurement system study is a simplified uncertainty analysis in which one assumes that only two sources of error are active and important, it makes sense that this RandR quantity is critical to the judgment of adequacy of the measurement device. The next step in the analysis is to compare the error to the required process capability. Specifically, one divides the RandR standard deviation by the tolerance width of the process to which this measurement device is applied. The concept revolves around the idea that the product has only a certain amount of acceptable variation that is allowable, and one would like to see how much of this interval is taken by the measurement error alone. Notice that this straightforward comparison does not directly show how the probability of error behaves, but it is still a useful indicator of this error.

The denominator in the measurement ratio computation can also be the standard deviation of the total process variation, which is calculated as the combination of both product variation and measurement error. Note that this computation will yield a different number if just the part variation is used instead of the total process variation. For example, a 30 percent ratio to the total process error will yield a higher ratio if just the part variation is used. If one computes the ratio and then inverts it, the index is very similar to the typical process capability indicator Cp. A ratio of 30 percent yields a Cp of 3.33, and a ratio of 10 percent yields a Cp of 10.0 when viewed in this fashion. From this perspective it can be seen that achieving a measurement system that has essentially zero impact on the process is quite difficult and roughly equivalent to finding a measurement process with a capability = 10.

PROBLEMS AND LIMITATIONS OF THE STANDARD MSA

Although the popularity of the MSA method has done much good in many industries by forcing people to study and improve their measurement systems, there are some problems and issues that arise that must be confronted when one tries to use this method generally. There are two broad categories of problems that are covered in detail in the subsequent chapters: weaknesses and pitfalls in the method as it exists and needed extensions of the standard method for more powerful analysis. Later chapters delve into these issues in detail, discussing the inherent variability in these estimates of the component variances and explaining how one can study more factors in the same MSA. There are also pitfalls in the application of these methods to cases in which there are only a few highly technical appraisers or operators.

The majority of the later chapters show how to extend the basic techniques to new situations that are not adequately covered in the standard approaches but are commonly encountered in modern industry and commerce. These situations include the addition of other measurement uncertainty sources into the analysis. This is important because many applications consider only the two sources of repeatability and reproducibility, whereas real measurement systems often have a multitude of important sources that are ignored by default. Perhaps the most important extensions of the basic technique will be to give viable ways in which to perform destructive and dynamic measurement systems. A large number of real measurement systems are currently being analyzed as if their values did not change with each repeat reading. In general, such a naive approach inflates the estimate of measurement error and can often lead to unnecessary purchases of new devices or other wasted efforts. In other cases the analysis of these destructive systems is simply not done because the available methods are not adequate, and this leads to many situations in which ignorance is certainly not bliss.

The final chapters deal with other critical topics that cry out for more coverage in practical situations. Attribute measurement systems are often far more numerous than variables systems, and yet the analyses that are usually offered are very crude and simplistic. There are benefits to be gained in providing effective methods for attribute measurement systems. There is also a need for closure in programs that affect the quality of measurement systems. Often, an MSA is neglected in company organizations and treated merely as a nuisance activity that must be endured for the sake of audits. Good methods for evaluating the impact of measurement systems quality on the bottom line are critical for justifying the work that is necessary to improve and maintain these systems. Once the decision to maintain these systems is made, one needs to know techniques that enable the updating of the systems over time.

Chapter 2 Take-Home Pay

1. Uncertainty analysis surveys all important sources of measurement error.

2. An MSA is a limited version of uncertainty analysis.

3. The standard MSA identifies repeatability, reproducibility, and parts effects.

4. The standard MSA does not fit all real measurement systems.

3

The Variability in the Estimate of Repeatability and Reproducibility

Often when an MSA is performed it is seen as a costly exercise that is necessary only because of the unrealistic demands of auditors or quality technicians. So there is a natural tendency among the personnel conducting the MSA study to shorten it and reduce the number of samples. Most of these people would probably admit that more samples are better than fewer samples, but there is little quantitative understanding of the actual impact. The cost of doing the study is clear, but the cost of doing a poor one is not as clear. This chapter seeks to develop intuition about the effect of sample size on the various elements of an MSA. Specifically, this first section will concentrate on repeatability, and later sections will develop reproducibility impacts.

EVALUATING THE VARIABILITY INHERENT IN THE REPEATABILITY ESTIMATES

An Initial Study

The standard MSA study introduced in Chapter 2 should contain 90 random independent results taken on 10 parts, with each measured three times by three appraisers. The measurements should be independent of one another; elaborate randomization and double blind schemes are often employed to ensure that this is true. Furthermore, the part values should be representative of the distribution of parts that is normally seen in the process in which the measurement systems are applied. Table 3.1 shows such a sample dataset.

The ANOVA method uses the data from such a study to estimate the variances, the standard deviations squared, and the important sources of variability including parts, appraisers, and equipment. The equipment variability is called repeatability and represents the variability in a measurement under homogeneous conditions. Table 2.2 showed the full details of such an ANOVA for the data presented in Table 3.1, complete with the three components of an MSA. This full analysis is the subject of a later section in this chapter. For now, consider a reanalysis of the data, estimating only parts and repeatability. Running the ANOVA with only one effect for parts and then using the residual as repeatability accomplishes this. The results of this analysis are shown in Table 3.1. An ANOVA is a standard technique of statistical analysis that has been successfully employed at least over the last half century. It partitions the total variation of a set of study data into independent components that add up to the whole. One component produced in every ANOVA application is an error or residual term that captures whatever cannot be explicitly traced to other changes.

It should be obvious that not including appraisers in the analysis, despite the fact that different appraisers were actually used in the study, distorts the results. This will be discussed in detail later. For now, concentrate on this reduced analysis and assume the repeatability standard deviation is estimated to be 0.85 mm. The question at hand is how would this estimate change if a different set of data from the same measurement system were collected and analyzed? This is an interesting question for at least two reasons. First, one might be interested in anticipating the results from the same measurement system if a new set of 90 measurements were to be taken. According to the model, the measurement errors would not likely be the same as in the first study, so the numerical results would almost certainly be quite different. Second, one might also be interested in how much variation there is in the estimates themselves. That is, how different could the results be based on the errors involved in getting the particular set of values that were produced in the study?

Table 3.1 ANOVA estimates of the standard MSA excluding appraisers.

Error Source	Variance Estimate	Standard Deviation	Percent Tolerance
Parts	7.15	2.67	—
Repeatability	0.64	0.80	24.0

A Replication of an MSA

There are theoretical answers to this question that can be found by examining Searle (1971) or other references, but a different approach will be taken here. One can simulate the variability in the results by generating multiple sets of data from the known or assumed properties of the measurement system (Rubenstein, 1981). For example, Table 3.2 shows another set of 90 numbers that were generated according to the same model as the data in Table 2.2. That is, the same normal distribution with the same inherent variability was used to draw new random measurement errors and produce new results. The true repeatability and part variations do not change, but the estimates of these quantities generated from the analysis do.

If one examines the individual elements of this table versus Table 2.2, there will be many small differences. And yet one can generally see that the patterns and sizes of the errors are similar between the two cases. Only a detailed analysis like the ANOVA MSA can summarize the apparent differences. Table 3.3 shows the ANOVA estimates of the repeatability and parts for this data.

Table 3.2 Another standard MSA dataset from an identical system.

Part	Operator A			Operator B			Operator C		
	Repeat 1	Repeat 2	Repeat 3	Repeat 1	Repeat 2	Repeat 3	Repeat 1	Repeat 2	Repeat 3
1	104.551	104.337	104.946	102.557	102.513	102.453	103.304	102.956	102.907
2	100.467	100.554	101.066	98.728	98.739	99.079	99.259	99.245	99.256
3	102.665	103.211	102.889	100.677	101.119	101.122	101.445	101.208	100.958
4	98.499	98.640	98.900	97.111	96.670	96.928	97.429	97.841	97.875
5	103.647	104.152	104.202	101.497	101.626	101.911	102.976	102.185	102.498
6	103.413	103.287	103.261	101.629	101.345	101.518	102.080	102.355	101.535
7	101.864	102.398	101.902	100.157	99.994	100.359	100.650	100.386	101.206
8	102.437	102.563	101.910	100.434	100.372	100.338	100.865	100.944	100.466
9	102.152	102.122	102.779	99.821	99.952	100.402	100.850	100.782	101.022
10	103.222	103.079	103.342	101.049	101.473	100.878	101.764	101.507	101.841

Table 3.3 The ANOVA estimates for another identical standard MSA.

Error Source	Variance Estimate	Standard Deviation	Percent Tolerance
Parts	2.53	1.59	—
Repeatability	0.86	0.93	28.0

A Collection of Replicated Studies

For veteran practitioners the level of change demonstrated by these two twin analyses may be expected, but for novices it may be a shock to see the estimates vary this much even though the fundamental system is identical. But this is commonplace for the estimation of variances. Of course, only two such trials do not do a very good job of addressing this issue. It would be better to run a number of such identical MSAs and look at the distribution of results. This is shown in Figure 3.1, in which the distribution of repeatability standard deviations for 25 independent simulated datasets is shown. Table 3.4 shows some of the summary statistics of this same set of 25 repeat MSAs. Figure 3.2 shows the histogram for the estimates of part variation, and Table 3.5 shows the summary statistics for part standard deviations.

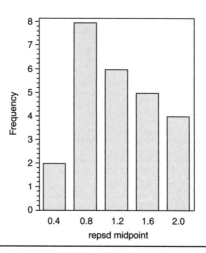

Figure 3.1   Variation in repeatability without appraiser analysis.

Table 3.4   Variation in repeatability without appraiser analysis.

N	25	Sum weights	25
Mean	1.22158505	Sum observations	30.5396263
Std deviation	0.46473779	Variance	0.21598121
Skewness	0.28685663	Kurtosis	0.4841231
Uncorrected SS	42.4903	Corrected SS	5.18354902
Coeff variation	38.0438337	Std error mean	0.09294756

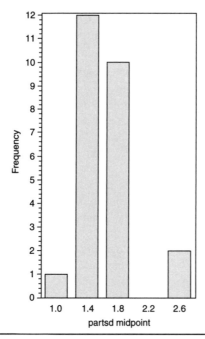

Figure 3.2 Variation in part estimates without appraiser analysis.

Table 3.5 Variation in part estimates without appraiser analysis.

N	25	Sum weights	25
Mean	1.64357408	Sum observations	41.089352
Std deviation	0.37079902	Variance	0.13749192
Skewness	1.16471285	Kurtosis	2.74586706
Uncorrected SS	70.8332	Corrected SS	3.29980598
Coeff variation	22.56053	Std error mean	0.0741598

Although there are many lessons to be learned from these results, there are at least three that are relevant to the further investigations to be presented:

1. The variability of repeatability standard deviations is high even with 90 results.

2. The effect of appraiser variability must be understood.

3. The part estimate variation appears to be relatively greater than the repeatability.

VARIABILITY IMPACT OF ADDING APPRAISERS TO THE ANOVA

Now it is time to revisit the full MSA with part, equipment, and appraiser components of variation. The same data that were generated for the previous repeatability demonstration can be reutilized by simply expanding the ANOVA to include parts and appraisers and still utilizing residuals for the repeatability component. The original analysis of the full set of data is presented here as Table 3.6. Table 3.7 shows the full results for the second scenario.

Figures 3.3 through 3.5 show the variability in the estimates of the standard deviations for parts, appraisers, and repeatability for 25 identical systems. Tables 3.8–3.10 show the summary statistics for these same 25 simulation runs.

It is clear that there is a great deal of variability in the estimates of the components of error for the recommended MSA. The appraiser and part effects seem to be especially sensitive, and the inclusion of reproducibility into the ANOVA definitely affects the estimates of repeatability and its consistency.

Table 3.6   Results of the ANOVA of the standard MSA study.

Error Source	Variance Estimate	Standard Deviation	Percent Tolerance
Parts	9.24	3.04	—
Appraiser	0.90	0.95	28.5
Repeatability	0.05	0.22	6.6
RandR	0.95	0.98	29.4

Table 3.7 Results of the ANOVA of the standard MSA study with appraisers.

Error Source	Variance Estimate	Standard Deviation	Percent Tolerance
Parts	9.33	2.44	—
Appraiser	0.95	1.09	16.3
Repeatability	0.07	0.92	13.8
RandR	1.02	1.01	30.3

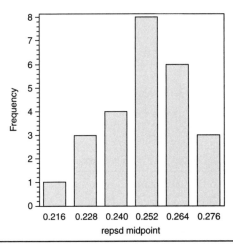

Figure 3.3 Variation in repeatability with appraiser analysis.

Table 3.8 Variation in repeatability with appraiser analysis.

N	25	Sum weights	25
Mean	0.25195497	Sum observations	6.29887424
Std deviation	0.01605632	Variance	0.00025781
Skewness	0.4289893	Kurtosis	−0.5550924
Uncorrected SS	1.59322	Corrected SS	0.00618733
Coeff variation	6.3726959	Std error mean	0.00321126

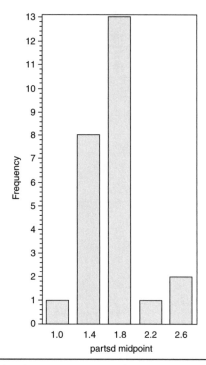

Figure 3.4 Variation in part estimates with appraiser analysis.

Table 3.9 Variation in part estimates with appraiser analysis.

N	25	Sum weights	25
Mean	1.69968343	Sum observations	42.4920858
Std deviation	0.36254344	Variance	0.13143775
Skewness	1.21339354	Kurtosis	2.61501897
Uncorrected SS	75.3776	Corrected SS	3.15450591
Coeff variation	21.3300568	Std error mean	0.07250869

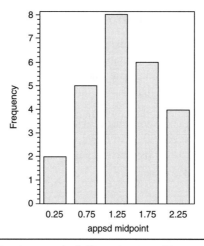

Figure 3.5   Variation in reproducibility estimates with appraiser analysis.

Table 3.10   Variation in reproducibility estimates with appraiser analysis.

N	25	Sum weights	25
Mean	1.37206017	Sum observations	34.3015042
Std deviation	0.55773248	Variance	0.31106552
Skewness	0.1907933	Kurtosis	–0.4084025
Uncorrected SS	54.5293	Corrected SS	7.46557239
Coeff variation	40.6492726	Std error mean	0.1115465

THE WHAT-IF SCENARIOS

With this baseline understanding of the inherent variability in the estimates from the standard MSA, it is possible to ask some fundamental questions about the impact of changing the basic MSA design. For example, one might be tempted to run two appraisers rather than one, or 15 parts rather than 10. Or one might want to cut down on the total number of repeats in order to save time, money, and effort in performing the study. The simulation approach will be used to examine these types of questions.

Changing the Number of Appraisers

Using the same basic measurement system performance as before, it is possible to generate 25 different simulations, pretending that the study contains only two appraisers at random from the population and thus has only 60 total measurements. Notice that this amounts to two changes from the basic setup, so any impact that is seen will be the outcome of the two mingled changes. Figures 3.6–3.8 show the distribution of the various estimates resulting from the ANOVA study of this new setup.

Tables 3.11–3.13 show the corresponding summary statistics for these same runs under the new scenario.

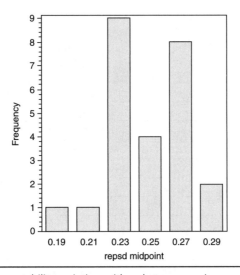

Figure 3.6 Repeatability variation with only two appraisers.

Table 3.11 Repeatability variation with only two appraisers.

N	25	Sum weights	25
Mean	0.24801064	Sum observations	6.20026612
Std deviation	0.02539521	Variance	0.00064492
Skewness	−0.2255128	Kurtosis	−0.6841088
Uncorrected SS	1.55321	Corrected SS	0.015478
Coeff variation	10.239565	Std error mean	0.00507904

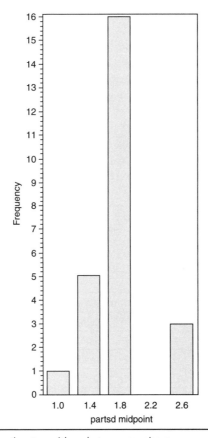

Figure 3.7   Part estimates with only two appraisers.

Table 3.12   Part variation with only two appraisers.

N	25	Sum weights	25
Mean	1.75887641	Sum observations	43.9719102
Std deviation	0.39606106	Variance	0.15686436
Skewness	1.02469935	Kurtosis	1.79114162
Uncorrected SS	81.1059	Corrected SS	3.76474467
Coeff variation	22.5178447	Std error mean	0.07921221

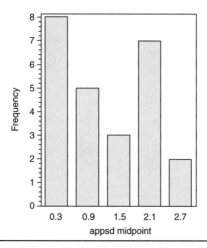

Figure 3.8   Reproducibility variation with only two appraisers.

Table 3.13   Reproducibility variation with only two appraisers.

N	25	Sum weights	25
Mean	1.36248907	Sum observations	34.0622268
Std deviation	0.81909982	Variance	0.67092451
Skewness	0.22436895	Kurtosis	−1.4805736
Uncorrected SS	62.5116	Corrected SS	16.1021883
Coeff variation	60.1179001	Std error mean	0.16381996

Notice that there are at least three things that happen when this attempt to reduce the cost of the standard MSA is attempted:

1. The variability in the reproducibility grows by a large amount.

2. The variability in the repeatability grows but not by as large an amount.

3. The variability in the parts estimates also grows but in an amount more similar to that displayed by the repeatability effect.

A tentative hypothesis (which can be confirmed by reference to the formulas in Searle, 1971) is that variability is controlled by the number of distinct elements used in the study. Low numbers of any element make them more sensitive to changes, and variability in repeatability affects all other components.

Changing the Number of Parts

Another common mutation in the standard MSA study is an increase in the number of parts. Often this increase in parts is coupled with a reduction in some other element, such as cutting from three appraisers to two or from three repeats to two. To keep things simple, consider only the impact of changing from 10 to 15 parts while retaining three appraisers and three repeats for a total of 135 measurements. The ANOVA remains the same in form, with effects estimated for parts, appraisers, and repeatability. The results of 25 simulation runs of this new scenario are shown in figures 3.9–3.11, and the corresponding summary statistics are shown in tables 3.14–3.16.

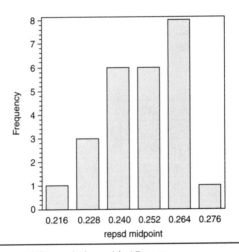

Figure 3.9 Repeatability variation with 15 parts.

Table 3.14 Repeatability variation with 15 parts.

N	25	Sum weights	25
Mean	0.24880406	Sum observations	6.22010158
Std deviation	0.01590788	Variance	0.00025306
Skewness	−0.1583617	Kurtosis	−0.9703772
Uncorrected SS	1.55366	Corrected SS	0.00607345
Coeff variation	6.39373746	Std error mean	0.00318158

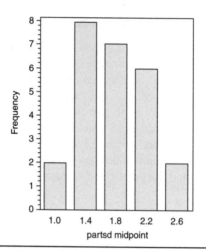

Figure 3.10   Part estimate variation with 15 parts.

Table 3.15   Part estimate variation with 15 parts.

N	25	Sum weights	25
Mean	1.78036784	Sum observations	44.5091959
Std deviation	0.46166181	Variance	0.21313163
Skewness	0.34760341	Kurtosis	−0.8241984
Uncorrected SS	84.3579	Corrected SS	5.11515905
Coeff variation	25.9306982	Std error mean	0.09233236

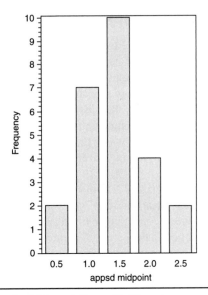

Figure 3.11 Reproducibility variation with 15 parts.

Table 3.16 Reproducibility variation with 15 parts.

N	25	Sum weights	25
Mean	1.40020197	Sum observations	35.0050493
Std deviation	0.56489162	Variance	0.31910254
Skewness	0.08714074	Kurtosis	−0.605188
Uncorrected SS	56.6726	Corrected SS	7.65846099
Coeff variation	40.3435812	Std error mean	0.11297832

The lesson here seems to confirm the hypotheses formulated in the previous section. All estimates seem to have improved in their variability, with parts gaining the most.

Changing the Number of Repeats

With the lessons taken from these two experiments, one might be tempted to cheat the system a little bit. Because the estimates seem to get better whenever the repeatability estimates get better (based on total number of repeats), perhaps one can run more repeats and in turn reduce the number of operators and parts accordingly. In most situations this would be the least expensive way to collect the 90 numbers for the MSA study. Consider the results of 25 runs of a scenario in which there are two appraisers, five parts, and 10 repeats for a total of 100 measurements (10 more than in the standard MSA design).

Figures 3.12–3.14 and tables 3.17–3.19 show the results of these simulations. Hopefully it is clear that this is not a good scenario. Whatever improvement was gained by including extra repeats seems more than overwhelmed by the reduction in the parts and appraiser elements.

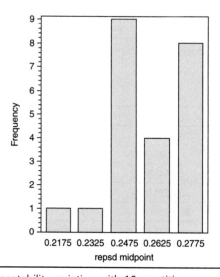

Figure 3.12   Repeatability variation with 10 repetitions.

Table 3.17   Repeatability variation with 10 repetitions.

N	25	Sum weights	25
Mean	0.24619345	Sum observations	6.15483613
Std deviation	0.01858504	Variance	0.0003454
Skewness	–0.1540858	Kurtosis	–0.3897923
Uncorrected SS	1.52357	Corrected SS	0.00828969
Coeff variation	7.54895764	Std error mean	0.00371701

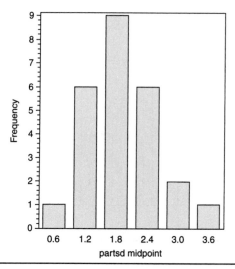

Figure 3.13 Part estimate variation with 10 repetitions.

Table 3.18 Part estimate variation with 10 repetitions.

N	25	Sum weights	25
Mean	1.89602004	Sum observations	47.4005011
Std deviation	0.73445274	Variance	0.53942083
Skewness	0.26344203	Kurtosis	−0.0909171
Uncorrected SS	102.8184	Corrected SS	12.9460999
Coeff variation	38.7365494	Std error mean	0.14689055

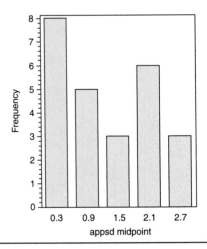

Figure 3.14 Reproducibility variation with 10 repetitions.

Table 3.19 Reproducibility variation with 10 repetitions.

N	25	Sum weights	25
Mean	1.34528653	Sum observations	33.6321633
Std deviation	0.81103596	Variance	0.65777932
Skewness	0.26586314	Kurtosis	−1.4144532
Uncorrected SS	61.0316	Corrected SS	15.7867038
Coeff variation	60.287228	Std error mean	0.16220719

THE IMPACT ON ESTIMATE OF THE RANDR INDEX

Although it has not been emphasized to this point in the text, the repeatability and reproducibility are often combined—that is, added—to form a single standard deviation representing the overall measurement error. Figure 3.15 shows the variability in this index for the 25 runs done previously for the standard 3×3×10 MSA design. Table 3.20 shows the numerical results from this analysis. Notice that there is quite a bit of variability in this RandR index and that it seems to be dominated by the variability in the reproducibility estimate. Once this fact is demonstrated it seems intuitive that this is true, because the sum generally should not have less variability than the worst of its components.

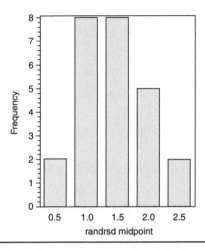

Figure 3.15   Variation in the RandR index estimate in the standard MSA.

Table 3.20   RandR variation.

N	25	Sum weights	25
Mean	1.40126474	Sum observations	35.0316186
Std deviation	0.54136971	Variance	0.29308116
Skewness	0.26668865	Kurtosis	–0.4709457
Uncorrected SS	56.12252	Corrected SS	7.03394788
Coeff variation	38.6343631	Std error mean	0.10827394

SUMMARY AND PRESCRIPTION

The standard MSA design does a pretty good job of estimating the repeatability estimate, a middling job with parts, and a poor job with appraisers. This is directly related to the number of elements recommended for each component. The use of only three appraisers is a real handicap to understanding the true effects of reproducibility. Ideally, as many as 15 appraisers should be used (if possible) to beef up this estimate. This is recommended even if one has to compensate by reducing repeats and parts. The estimate of part variation will suffer, but because this is not the chief purpose of an MSA study, it could be a wise trade-off when it is necessary.

Chapter 3 Take-Home Pay

1. The estimates of the error sources have inherent variation.

2. The standard MSA provides a reasonably consistent estimate of repeatability.

3. The standard MSA provides a far less consistent estimate of reproducibility.

4. The standard MSA provides a less consistent estimate of part variability.

5. A consistent estimate of reproducibility requires as many as 15 appraisers.

6. The estimate of RandR is dominated by the uncertainty of reproducibility.

4

Extending the Standard MSA to Other Sources of Uncertainty

EXTENDING THE ANOVA MSA TO MORE SOURCES

There are two threads, developed in earlier chapters, that can be woven together in this chapter to make a more powerful MSA. The first thread was developed in Chapter 1, in which the impacts of measurement system error were shown to have potentially costly effects on product sorting, process control, and many other activities. So even though one chooses to study only repeatability and reproducibility, it does not mean that other sources of error do not exist. And if they do exist, it is likely that they can cause errors and mistakes. The conclusion from this line of reasoning is that it would be a good thing to understand these other sources of uncertainty and to manage them.

The second thread was developed in Chapter 3, in which it was demonstrated that the ANOVA can accommodate changes to the MSA design and can actually benefit from these modifications by producing more precise estimates. These two separate threads lead one to think that there might be even more advantage in using an ANOVA to study a large set of uncertainty sources. Much of this chapter will be devoted to demonstrating and detailing this power.

Systems in Which Appraisers Are Not a Dominant Effect

Many measurement systems used in real situations do not require an operator or an appraiser. These kinds of measurement systems include laser systems, automatic laboratory equipment, and computer-monitored devices. In other situations there might be a small appraiser effect, but it is not dominant. In either case the standard MSA recommendation for including three appraisers in the study does not make much sense. In these cases there are at least two possibilities: study only repeatability or study an additional uncertainty source in place of an appraiser.

For example, an automated laser system might not need an appraiser during normal operation after a correct setup procedure has been performed. Therefore, the operator or appraiser may have no effect on operations but could have an enormous indirect effect if he or she performs an inadequate setup procedure. The standard MSA study might then be repeated three times under three different setups. These three setups might be performed by different people or even by the same person. One might run 10 parts with three repeat measurements under each of the three setups for the usual total of 90 measurements. All analysis could proceed exactly as before, as an ANOVA with parts and setup periods with repeatability given by residual. Only the interpretation would be different, with reproducibility now representing the effect of setup and not the appraiser. Table 4.1 shows the data table as it might appear for such a modified MSA, and Table 4.2 shows the ANOVA result table splitting out the three components, with setup replacing appraiser. Note that sample size effects will weigh most heavily on the setup term now because it has only three elements.

Table 4.1 Data details for the standard MSA study.

	Setup One			Setup Two			Setup Three		
Part	Repeat 1	Repeat 2	Repeat 3	Repeat 1	Repeat 2	Repeat 3	Repeat 1	Repeat 2	Repeat 3
1	56.96	57.27	67.10	55.91	55.91	55.90	56.04	56.25	56.05
2	54.40	54.47	54.38	53.25	53.47	53.49	53.45	53.38	53.53
3	44.92	44.90	44.41	43.87	43.74	43.55	43.86	44.08	43.92
4	39.68	39.77	39.73	38.86	38.68	38.93	38.91	38.86	38.95
5	51.42	51.45	51.43	50.15	50.12	50.00	50.09	50.38	50.32
6	50.75	50.65	51.00	49.94	49.68	49.96	49.92	50.31	49.94
7	45.23	45.07	45.16	44.10	44.06	44.05	44.39	44.31	44.36
8	49.35	49.44	49.42	48.20	48.15	47.85	48.34	48.30	48.23
9	55.53	55.27	55.39	54.07	54.46	54.31	54.63	54.60	54.47
10	53.15	53.28	53.27	52.02	52.39	52.36	52.34	52.38	52.02

Table 4.2 Results of replacing appraiser with setup in the MSA study.

Error Source	Variance Estimate	Standard Deviation	Percent Tolerance
Parts	29.51	5.43	—
Setup	0.35	0.59	35.4
Repeatability	0.14	0.22	8.4
RandR	0.37	0.61	36.6

Extending the Analysis to Interactions

It is also possible to improve an MSA without physically changing anything in the design. This is because the standard MSA study not only allows an analysis of the effects of each component of variation separately but also permits the examination of effects that depend on two or more sources simultaneously. These effects are called *interactions* (Ott, 1984). An interaction between an appraiser and a part could stem from a difference in the way some appraisers deal with certain types of products. A strong appraiser might easily handle a heavy, large part, whereas a weaker appraiser would have more trouble. Or, a small operator might do better with a delicate measurement task, whereas a larger person might crush or deform the product. If there were two appraisers and two parts, a strong interaction would occur when the situation was as shown in Figure 4.1.

When there are three appraisers and 10 parts the interaction is not a single effect but rather a combination of effects across all pairs of appraisers and parts. If the measurement variation can be adequately explained as a simple summation of the separate effects of part and appraiser, the interaction effect will be near zero. Figure 4.2 shows an example of this more complex interaction effect.

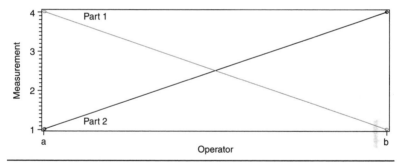

Figure 4.1 Example of simple interaction.

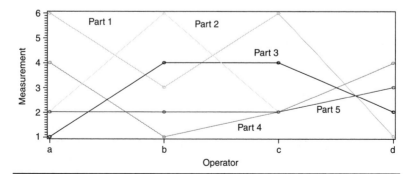

Figure 4.2 Example of complex interaction.

The example data used in Chapter 2 (Table 2.1) are modified to include an interaction effect and are shown in Table 4.3.

Now the ANOVA may be computed with one new term, the interaction between parts and appraisers. Table 4.4 shows these results.

There are two important things to note about this new breakdown of the data with an additional component. First, there is another effect that has been identified. If this interaction component is large, one can specifically look to certain features of the measurement system in hopes of making improvements. This can be a tremendous advantage when the measurement system is inadequate. If the results are as presented in Table 4.4, the interaction is large and should be remedied, perhaps by providing help with large objects.

Second, as important as the advantage of greater understanding can be, there is another impact that can be just as valuable. To understand this advantage, consider Table 4.5, in which the same data are reanalyzed without the identification of the interaction effect.

Table 4.3   Data details for the standard with interaction MSA study.

Part	Appraiser A			Appraiser B			Appraiser C		
	Repeat 1	Repeat 2	Repeat 3	Repeat 1	Repeat 2	Repeat 3	Repeat 1	Repeat 2	Repeat 3
1	102.408	102.924	102.633	101.226	101.214	101.201	102.342	102.695	102.363
2	102.629	102.743	102.598	98.125	98.487	98.528	101.212	101.099	101.350
3	98.103	98.077	97.263	93.766	93.546	93.228	96.518	96.874	95.616
4	95.137	95.281	95.215	91.174	90.871	91.287	94.017	93.929	94.082
5	101.301	101.340	101.318	96.589	96.536	96.340	99.239	99.729	99.629
6	100.560	100.399	100.983	96.630	96.189	96.642	99.339	99.985	99.375
7	98.000	97.720	97.881	93.517	93.449	93.434	96.767	96.622	96.714
8	100.305	100.454	100.414	95.785	95.709	95.215	98.781	98.717	98.603
9	103.235	102.793	102.989	98.198	98.864	98.597	101.888	101.846	101.635
10	101.937	102.142	102.135	97.444	97.898	98.015	100.745	100.815	100.210

Table 4.4   Results of the standard with interaction MSA study.

Error Source	Variance Estimate	Standard Deviation	Percent Tolerance
Parts	2.78	1.67	—
Appraiser	4.41	2.10	64.0
Appraiser*Part	0.23	0.48	14.4
Repeatability	0.05	0.22	6.6
RandR	4.64	2.15	64.5

Compare tables 4.4 and 4.5, and it is clear that the interaction effect must have originally been hidden in the repeatability term. Allowing the ANOVA to identify and estimate this interaction term is equivalent to applying a filter that removes the interaction effect from the repeatability estimate. Including an interaction term in the analysis can be a very valuable filter.

There is an additional benefit that can come from this separation of the repeatability and interaction effects. In Chapter 3 it was shown that the variability of the different error components depends on the number of elements in the study and the size of the repeatability term. Thus, the repeatability term can be decreased by this separation, and this, in turn, can decrease the variability of the estimates. As a demonstration, figures 4.3 and 4.4 show

Table 4.5 Results of ignoring the interaction effect in the MSA study.

Error Source	Variance Estimate	Standard Deviation	Percent Tolerance
Parts	7.34	2.71	—
Appraiser	4.43	2.10	63.0
Repeatability	0.21	0.46	13.8
RandR	4.64	2.15	64.5

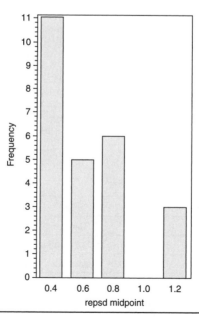

Figure 4.3 Variation in repeatability with interaction.

the results of 25 different MSAs with and without interaction removed for repeatability. These results, shown in tables 4.6 and 4.7, clearly demonstrate the possible impact of including a strong interaction in an MSA study.

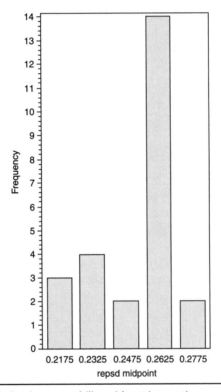

Figure 4.4   Variation in repeatability without interaction.

Table 4.6   Summary statistics with interaction.

N	25	Sum weights	25
Mean	0.63254976	Sum observations	15.8137439
Std deviation	0.28861163	Variance	0.08329667
Skewness	0.88912439	Kurtosis	−0.0430122
Uncorrected SS	12.0021	Corrected SS	1.99912016
Coeff variation	45.6267082	Std error mean	0.05772233

Table 4.7 Summary statistics without interaction.

N	25	Sum weights	25
Mean	0.25178441	Sum observations	6.29461031
Std deviation	0.01803381	Variance	0.00032522
Skewness	−0.6383985	Kurtosis	−0.8004366
Uncorrected SS	1.59269	Corrected SS	0.00780524
Coeff variation	7.16240193	Std error mean	0.00360676

Adding New Sources

Once one is aware that an MSA can be extended to other effects, which is valuable both in providing new information and in improving the existing estimates, one should be quite eager to use this knowledge. Perhaps the initial uncertainty analysis shows that there are four chief sources of measurement error: appraisers, setup, position, and repeatability. More pointers on possible designs for MSA studies will be discussed in Chapter 5, but for now one can use the standard MSA as a template and merely extend it to cover the new sources. That is, one can simply take the standard appraiser × part × repeat scenario and duplicate over different setups and positions. If there are two setups and two positions chosen for the analysis, one might choose four parts, three appraisers, and two repeats to total $2 \times 2 \times 4 \times 3 \times 2 = 96$ measurements. Note the implications that the small sample sizes have on the resulting estimates. Table 4.8 shows the possible data from such a study, and Table 4.9 shows the analysis of this data. Table 4.10 shows, for comparison's sake, the results if the same data are analyzed only for effects of appraiser and repeatability.

Table 4.8 Extending an MSA to other sources.

Obs	Seq	Part	App	Setup	Position	Meas
1	1	1	1	1	1	1005.46
2	1	1	1	1	1	1005.97
3	1	1	1	1	2	1004.10
4	1	1	1	1	2	1003.89
5	1	1	1	2	1	1002.83
6	1	1	1	2	1	1002.82
7	1	1	1	2	2	1001.36

(continued)

Table 4.8   Extending an MSA to other sources.

Obs	Seq	Part	App	Setup	Position	Meas
8	1	1	1	2	2	1001.71
9	1	1	2	1	1	1004.01
10	1	1	2	1	1	1004.16
11	1	1	2	1	2	1002.70
12	1	1	2	1	2	1002.55
13	1	1	2	2	1	1001.37
14	1	1	2	2	1	1001.74
15	1	1	2	2	2	1000.20
16	1	1	2	2	2	1000.02
17	1	1	3	1	1	1004.21
18	1	1	3	1	1	1004.46
19	1	1	3	1	2	1002.99
20	1	1	3	1	2	1002.96
21	1	1	3	2	1	1001.10
22	1	1	3	2	1	1001.94
23	1	1	3	2	2	1000.15
24	1	1	3	2	2	999.83
25	1	2	1	1	1	1004.57
26	1	2	1	1	1	1004.93
27	1	2	1	1	2	1003.09
28	1	2	1	1	2	1002.61
29	1	2	1	2	1	1001.70
30	1	2	1	2	1	1001.63
31	1	2	1	2	2	1000.36
32	1	2	1	2	2	1000.06
33	1	2	2	1	1	1003.10
34	1	2	2	1	1	1002.97
35	1	2	2	1	2	1001.30
36	1	2	2	1	2	1001.45
37	1	2	2	2	1	1000.40
38	1	2	2	2	1	1000.43
39	1	2	2	2	2	998.84
40	1	2	2	2	2	998.45

(continued)

Table 4.8 Extending an MSA to other sources.

Obs	Seq	Part	App	Setup	Position	Meas
41	1	2	3	1	1	1002.70
42	1	2	3	1	1	1002.51
43	1	2	3	1	2	1000.96
44	1	2	3	1	2	1001.45
45	1	2	3	2	1	1000.30
46	1	2	3	2	1	999.92
47	1	2	3	2	2	998.18
48	1	2	3	2	2	998.76
49	1	3	1	1	1	999.61
50	1	3	1	1	1	999.16
51	1	3	1	1	2	998.04
52	1	3	1	1	2	997.87
53	1	3	1	2	1	997.466
54	1	3	1	2	1	996.857
55	1	3	1	2	2	995.268
56	1	3	1	2	2	994.988
57	1	3	2	1	1	997.777
58	1	3	2	1	1	997.760
59	1	3	2	1	2	996.115
60	1	3	2	1	2	996.100
61	1	3	2	2	1	995.515
62	1	3	2	2	1	995.370
63	1	3	2	2	2	993.885
64	1	3	2	2	2	994.043
65	1	3	3	1	1	998.493
66	1	3	3	1	1	998.454
67	1	3	3	1	2	996.594
68	1	3	3	1	2	996.518
69	1	3	3	2	1	994.974
70	1	3	3	2	1	995.672
71	1	3	3	2	2	994.032
72	1	3	3	2	2	993.918
73	1	4	1	1	1	997.123

(continued)

(continued)

Table 4.8   Extending an MSA to other sources.

Obs	Seq	Part	App	Setup	Position	Meas
74	1	4	1	1	1	996.680
75	1	4	1	1	2	995.300
76	1	4	1	1	2	994.855
77	1	4	1	2	1	994.470
78	1	4	1	2	1	994.204
79	1	4	1	2	2	993.051
80	1	4	1	2	2	993.009
81	1	4	2	1	1	995.426
82	1	4	2	1	1	995.391
83	1	4	2	1	2	994.020
84	1	4	2	1	2	994.013
85	1	4	2	2	1	992.617
86	1	4	2	2	1	993.072
87	1	4	2	2	2	991.612
88	1	4	2	2	2	991.475
89	1	4	3	1	1	995.846
90	1	4	3	1	1	995.241
91	1	4	3	1	2	993.980
92	1	4	3	1	2	993.766
93	1	4	3	2	1	993.324
94	1	4	3	2	1	992.968
95	1	4	3	2	2	991.348
96	1	4	3	2	2	991.288

Table 4.9   Analysis of other sources.

Error Source	Variance Estimate	Standard Deviation	Percent Tolerance
Parts	16.59	4.07	—
Appraiser	0.71	0.84	12.6
Setup	3.37	1.84	27.6
Position	1.24	1.11	16.6
Repeatability	0.06	0.24	3.6
Uncertainty	5.38	2.32	34.8

Table 4.10 Results of analysis with other sources ignored.

Error Source	Variance Estimate	Standard Deviation	Percent Tolerance
Parts	16.49	4.06	—
Appraiser	0.63	0.79	11.8
Repeatability	2.52	1.59	23.8
RandR	3.15	1.77	26.5

Continuous Effects

Sometimes there are effects that impact measurement error that are not discrete, such as observations of ambient temperature, relative humidity, and elapsed time. Fortunately, an MSA can be easily extended to allow the study of these continuous sources of variation through covariance analysis (Snedecor, 1967). *Covariance analysis* is a modification of the basic ANOVA that allows one to correct the analysis for background effects. One might correctly suspect that removal of these kinds of effects will again have the double positive impact of adding to the understanding of the performance of the system and improving the variability of the estimates of all effects. Consider Table 4.11, which shows the possible data from a study in which temperature is allowed to fluctuate in a random fashion throughout the running of the standard study.

Table 4.11 An MSA with a continuous source.

Obs	Seq	Part	App	Temp	Meas
1	1	1	1	86.2584	1089.97
2	1	1	1	80.5448	1084.77
3	1	1	1	88.3268	1092.27
4	1	1	2	81.4188	1083.57
5	1	1	2	88.7071	1090.84
6	1	1	2	86.5392	1088.66
7	1	1	3	80.3611	1082.70
8	1	1	3	88.3684	1091.06
9	1	1	3	82.5747	1084.94
10	1	2	1	88.0134	1090.64
11	1	2	1	87.2932	1090.04
12	1	2	1	87.1015	1089.70

(continued)

Table 4.11 An MSA with a continuous source.

Obs	Seq	Part	App	Temp	Meas
13	1	2	2	84.4059	1085.30
14	1	2	2	88.6871	1089.94
15	1	2	2	88.5547	1089.85
16	1	2	3	84.1781	1085.39
17	1	2	3	88.0553	1089.15
18	1	2	3	81.7830	1083.13
19	1	3	1	86.7369	1084.84
20	1	3	1	82.9302	1081.01
21	1	3	1	81.0217	1078.28
22	1	3	2	81.3430	1077.88
23	1	3	2	86.9190	1083.23
24	1	3	2	88.3104	1084.31
25	1	3	3	87.2513	1083.77
26	1	3	3	86.0508	1082.92
27	1	3	3	88.1847	1084.80
28	1	4	1	82.8463	1077.98
29	1	4	1	84.0869	1079.37
30	1	4	1	81.8119	1077.03
31	1	4	2	86.4686	1080.41
32	1	4	2	84.8165	1078.46
33	1	4	2	87.5324	1081.59
34	1	4	3	80.8496	1074.87
35	1	4	3	83.3189	1077.25
36	1	4	3	81.8275	1075.91
37	1	5	1	86.6754	1087.98
38	1	5	1	80.5450	1081.88
39	1	5	1	86.4517	1087.77
40	1	5	2	80.3922	1079.75
41	1	5	2	88.9663	1088.27
42	1	5	2	89.1068	1088.22
43	1	5	3	81.9819	1081.22
44	1	5	3	87.4961	1087.23
45	1	5	3	80.9543	1080.58

(continued)

Table 4.11　An MSA with a continuous source.

Obs	Seq	Part	App	Temp	Meas
46	1	6	1	80.0563	1080.62
47	1	6	1	82.0179	1082.42
48	1	6	1	84.4602	1085.44
49	1	6	2	85.4037	1084.80
50	1	6	2	81.6287	1080.59
51	1	6	2	82.1322	1081.54
52	1	6	3	89.4229	1088.76
53	1	6	3	81.4308	1081.42
54	1	6	3	80.8638	1080.24
55	1	7	1	80.7717	1078.77
56	1	7	1	87.3809	1085.10
57	1	7	1	86.0182	1083.90
58	1	7	2	87.3168	1083.60
59	1	7	2	80.9405	1077.16
60	1	7	2	81.6101	1077.81
61	1	7	3	84.7669	1081.53
62	1	7	3	82.6574	1079.28
63	1	7	3	88.2554	1084.97
64	1	8	1	87.9050	1088.21
65	1	8	1	84.3134	1084.77
66	1	8	1	80.1846	1080.60
67	1	8	2	87.7789	1086.33
68	1	8	2	85.7184	1084.20
69	1	8	2	87.1445	1085.13
70	1	8	3	84.8214	1083.60
71	1	8	3	80.4869	1079.20
72	1	8	3	86.6659	1085.27
73	1	9	1	87.1839	1090.42
74	1	9	1	86.3509	1089.14
75	1	9	1	81.3040	1084.29
76	1	9	2	84.6792	1085.65
77	1	9	2	88.7365	1090.37
78	1	9	2	85.7091	1087.08

(continued)

(continued)

Table 4.11 An MSA with a continuous source.

Obs	Seq	Part	App	Temp	Meas
79	1	9	3	83.9176	1085.81
80	1	9	3	86.9138	1088.76
81	1	9	3	83.9973	1085.63
82	1	10	1	82.4087	1084.35
83	1	10	1	86.6667	1088.81
84	1	10	1	85.5908	1087.73
85	1	10	2	86.3033	1086.52
86	1	10	2	88.0791	1088.75
87	1	10	2	86.0919	1086.88
88	1	10	3	89.5215	1090.27
89	1	10	3	80.7657	1081.58
90	1	10	3	89.9056	1090.12

Tables 4.12 and 4.13 show the results of the analysis of this data with and without the removal of continuous temperature effect.

Table 4.12 An MSA including the continuous source.

Error Source	Variance Estimate	Standard Deviation	Percent Tolerance
Parts	7.21	2.69	—
Appraiser	0.78	0.88	13.2
Repeatability	0.06	0.24	3.6
RandR	0.84	0.92	13.8

Table 4.13 Analysis ignoring the continuous source.

Error Source	Variance Estimate	Standard Deviation	Percent Tolerance
Parts	9.90	3.15	—
Appraiser	0.11	0.33	4.0
Repeatability	8.49	2.91	43.6
RandR	8.60	2.93	43.9

SUMMARY

The standard MSA study should be just the starting point for powerful studies that identify and isolate more sources of measurement uncertainty. The choice of the sources to use is still a matter of economics and feasibility, but one should not be limited to just two sources. Once these sources have been identified, there are still technical choices as to the number of elements of each source, the number of samples, and the actual pattern of the trials to utilize. These details are the subject of the next chapter.

Chapter 4 Take-Home Pay

1. Appraisers can be replaced by other sources such as setup.

2. Interactions can be added to the analysis without additional data collection.

3. More sources such as setup, time, machine, and so on can be added to an MSA.

4. Continuous effects such as temperature can be added to an MSA.

5. More sources help one to better understand the measurement system.

6. Generally, more sources improve the consistency of all estimates.

5

Deformative Measurements

COMPLEX MEASUREMENT SYSTEMS

The AIAG MSA approach is primarily targeted at systems in which the parts are stable and there is no change between repetitions. But most systems do cause some change from repetition to repetition. For example, a caliper might be used on a soft rubber sample to measure thickness. Unless one waits an extremely long time to allow the rubber to regain its original width, there will be a depression maintained at the location of the measurement. If the repeat measurement is made at the same location, it is likely that the true width at that location will be smaller. So differences between repeated measurements conducted by identical operators and identical parts will be inflated by this change in the part's true value (Bergeret, 2001). These kinds of situations are labeled nonrepeatable MSAs and are shown in Figure 5.1. An example set of data is shown in Table 5.1. It will be necessary

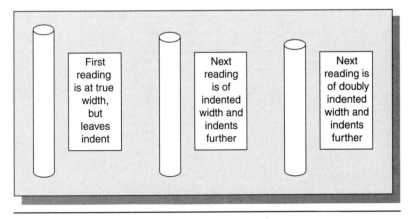

Figure 5.1   Nonrepeatable measurement system diagram.

Table 5.1 Example of data from a nonrepeatable systems diagram.

Part	Repeat	Appraiser	Width	Repeat	Appraiser	Width	Repeat	Appraiser	Width
I	1	A	100.8	2	B	97.8	3	C	96.9
I	4	A	97.3	5	B	94.5	6	C	93.3
I	7	A	93.5	8	B	92.4	9	C	91.8
II	1	A	102.1	2	B	100.6	3	C	100.6
II	4	A	98.0	5	B	97.0	6	C	94.5
II	7	A	97.0	8	B	96.4	9	C	92.2
III	1	A	97.8	2	B	94.5	3	C	95.5
III	4	A	5	23	B	91.5	6	C	90.3
III	7	A	93.6	8	B	90.0	9	C	88.9
IV	1	A	97.5	2	B	93.0	3	C	91.7
IV	4	A	93.2	5	B	92.1	6	C	90.5
IV	7	A	92.6	8	B	87.6	9	C	85.2
V	1	A	98.7	2	B	96.5	3	C	93.1
V	4	A	95.5	5	B	93.0	6	C	91.0
V	7	A	91.3	8	B	89.8	9	C	87.7
VI	1	A	98.4	2	B	96.6	3	C	94.5
VI	4	A	95.2	5	B	93.2	6	C	92.0
VI	7	A	91.6	8	B	89.4	9	C	87.4
VII	1	A	103.6	2	B	100.4	3	C	98.8
VII	4	A	100.6	5	B	98.5	6	C	96.1
VII	7	A	97.8	8	B	95.0	9	C	93.8
VIII	1	A	99.3	2	B	96.0	3	C	93.3
VIII	4	A	96.9	5	B	93.3	6	C	92.1
VIII	7	A	93.2	8	B	91.1	9	C	89.5
IX	1	A	100.0	2	B	96.3	3	C	95.0
IX	4	A	96.8	5	B	96.7	6	C	92.7
IX	7	A	93.7	8	B	92.8	9	C	89.9
X	1	A	95.2	2	B	91.9	3	C	92.1
X	4	A	92.7	5	B	88.6	6	C	89.4
X	7	A	90.9	8	B	87.5	9	C	87.0

to perform an appropriate analysis to separate the hidden part variation from the true repeatability to get correct assessments of the different measurement error components.

Another more extreme scenario is the situation in which the measurement destroys the sample or part completely. This could be through incineration (as in a thermodynamical study) a change in chemical composition, or elongation past the breaking point of the material. In these cases there is no possible way in which to get a repeat of the same part under identical conditions. Notice that this destruction can be undetectable to the eye and might only be detectable by sensor. This more extreme situation is labeled a destructive MSA and is illustrated by Figure 5.2 and the data in Table 5.2.

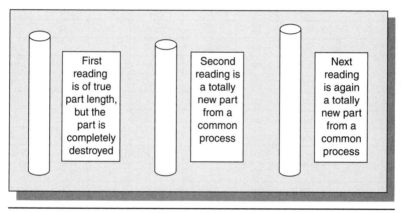

Figure 5.2 Destructive measurement system example.

Table 5.2 Data for a destructive example.

Repeat	Appraiser	Measure	Repeat	Appraiser	Measure	Repeat	Appraiser	Measure
1	A	101.1	2	B	99.8	3	C	102.0
4	A	105.3	5	B	103.1	6	C	99.3
7	A	95.6	8	B	92.4	9	C	94.4
10	A	103.1	11	B	104.0	12	C	106.6
13	A	104.8	14	B	102.0	15	C	96.2
16	A	97.0	17	B	98.8	18	C	101.2
19	A	99.6	20	B	98.5	21	C	100.6

(continued)

(continued)

Table 5.2   Data for a destructive example.

Repeat	Appraiser	Measure	Repeat	Appraiser	Measure	Repeat	Appraiser	Measure
22	A	99.8	23	B	93.0	24	C	90.3
25	A	95.6	26	B	98.0	27	C	104.3
28	A	99.5	29	B	96.4	30	C	94.7
31	A	94.4	32	B	92.1	33	C	92.3
34	A	99.6	35	B	101.3	36	C	103.2
37	A	100.4	38	B	98.5	39	C	94.0
40	A	95.5	41	B	94.4	42	C	97.0
43	A	103.3	44	B	105.8	45	C	103.1
46	A	99.4	47	B	97.2	48	C	94.5
49	A	96.4	50	B	98.2	51	C	102.3
52	A	105.6	53	B	103.0	54	C	96.4
55	A	103.9	56	B	100.4	57	C	99.6
58	A	104.6	59	B	107.0	60	C	108.1
61	A	109.7	62	B	103.0	63	C	96.4
64	A	99.3	65	B	96.6	66	C	96.3
67	A	103.8	68	B	103.3	69	C	102.4
70	A	100.2	71	B	93.4	72	C	89.5
73	A	100.3	74	B	98.3	75	C	100.2
76	A	104.8	77	B	105.3	78	C	98.7
79	A	95.7	80	B	92.8	81	C	92.6
82	A	96.2	83	B	95.3	84	C	98.1
85	A	99.5	86	B	93.6	87	C	91.1
88	A	90.9	89	B	89.9	90	C	96.0

THE NAIVE APPROACH TO NONREPEATABLE MSAS

There are at least three possible ways to approach an MSA in a situation in which the repeats contain some part variation even though they do not result in the complete destruction of the part. One possibility is to pretend that there is nothing nonrepetitive about the situation and proceed with the regular analysis. The estimate of repeatability will be contaminated—in fact, inflated—by the embedded part variations. But if the estimate is still precise enough and still good enough for the application, this inflated number does not change one's decision. One can use the instrument without prejudice, knowing that the real repeatability error is guaranteed to be less than that evaluated in the naive MSA. For most practical studies this naive approach is a good first option.

Consider the first example, given as Table 5.1. There is an embedded change in the true part values, but one can choose to ignore it and perform a standard MSA. If one pursues this approach, one might get the results shown in Table 5.3, which provides the resulting estimates of the three variances, their standard deviations, and the percent tolerances using a tolerance range of +/–40 mm.

In this example the system still achieves an overall RandR index of less than 30 percent, so it might be considered marginally acceptable even though it is likely that the actual measurement error is less than the estimate. Notice, however, that this measurement system would be considered only marginally acceptable by AIAG standards. If a better, less naive estimate of the true measurement error that separated the part variation were available, the systems might achieve full acceptability without any additional improvement actions.

Table 5.3 The naive MSA approach to a nonrepetitive process.

Error Source	Variance Estimate	Standard Deviation	6*Std Dev	100*6*SD/tol range
Parts	5.29	2.30	—	—
Appraiser	4.11	2.03	12.16	15.2%
Repeatability	6.84	2.62	15.69	19.6%
RandR	10.95	3.31	19.85	24.8%

A CORRECTIVE APPROACH TO NONREPEATABLE MSAS

Another approach is to design the study so that the part-to-part variation is minimized in some way. This may be difficult to do, but it is usually possible for nonrepeatable but nondestructive measurements. For example, in the case of the caliper effects that form the basis for the same example, one could wait the required time between measurements to allow the material to regain its original width. With rubber samples this could take 10 minutes between readings. If the measurement is a temperature, the conditioning time might be in hours. And if the conditioning is some form of aging, it may take days or even years to achieve the proper conditions that will eliminate the part-to-part variation from contaminating the repeatability estimate.

Sometimes expert opinion of physical models can lead one to the correct and adequate time interval to condition the samples in this way, but more often it must be estimated from studies done with the same measuring equipment with varying time intervals. Here is an example of how such a study might progress for the caliper example. Essentially, a partial MSA is performed at various increasing time intervals until no significant difference in the estimates is found. Even with this study, however, one must be conscious of the need for adequate sample size to see real differences. Table 5.4 illustrates an example of such an attempt to find an adequate relaxation time for the caliper nonrepetitive measurement system. All the entries in the table are variance estimates. It would appear that using a minimum relaxation time of between 10 and 20 minutes would be sufficient to damp out most of the contamination from embedded part variation. Of course, this would also mean that the time to conduct the MSA study would be at least 10 times longer if one minute were the original plan.

Table 5.4   Example of finding an adequate recovery time.

Relaxation Time	Part	Appraiser	Repeatability	RandR
1 minute	5.29	4.11	6.84	10.95
2 minutes	5.40	3.05	5.12	8.17
5 minutes	5.03	2.85	4.60	7.45
10 minutes	5.23	2.50	3.29	5.79
20 minutes	5.25	2.45	3.12	5.67
30 minutes	5.17	2.48	3.10	5.58
40 minutes	5.20	2.46	3.10	5.56

A DESIGN APPROACH TO NONREPEATABLE MSAS

An alternate way of changing the design so that it minimizes or eliminates the contamination of the repeatability error estimate is to enlarge the study so that the pattern of change that occurs is exactly repeated more than once. Again using the caliper example, one might assume that the pattern of change depends only on the number of repetitions if the time between repetitions is constant. For example, one might assume that repetition 1 is unaffected, repetition 2 drops by a value of delta1, repetition 3 by delta2, and so on in a trend pattern that is independent of the actual product width or the appraiser.

If this is true, the MSA design might be expanded to use this assumption to an advantage in decontaminating the repeatability estimate.

Consider the design in which the same appraiser measures the same part two times and then waits an appropriate time for relaxation, perhaps 30 minutes, before the same appraiser makes two more repeats on the same part. Similar actions are made for other parts and other appraisers. As a result of this effort there are several sets of repeats that can be used to estimate true measurement effects with reduced contamination by part changes. To see how this might work, consider the four repeats made by appraiser A on part 1. The true part values are assumed to be the same between time periods and to be changed in the same fashion and by the same delta between repeats 1 and 2 as between repeats 3 and 4. So the computed difference between repeats 1 and 2 reduces to an estimate of that delta plus two inherent repeatability errors. And the same is true for the difference between repeats 3 and 4. So if one further computes the second-level difference between these two first-level differences, it should mitigate the deltas, subtract them out, and estimate the combination of four independent repetition errors. Thus, if one reduces the original data to these final differences, one can compute an estimate of the MSA quantities as required. The results are shown in Table 5.5.

The standard deviation of the second differences is 0.50.

Table 5.5 The two-time-period approach.

Part	Appraiser	Rep 1	Rep 2	Dif 1	Rep 3	Rep 4	Dif 1	Dif 2
1	A	101.8	96.3	5.5	100.5	95.8	4.7	0.8
2	A	102.0	99.0	3.0	103.5	99.4	4.1	−1.1
3	A	100.6	94.0	6.6	97.7	95.2	2.5	4.1
4	A	99.2	94.0	5.2	97.8	93.3	4.5	0.7
5	A	98.5	93.0	5.5	99.4	94.3	5.1	0.4
6	A	100.2	95.1	5.1	100.9	95.0	5.8	−0.7
7	A	103.7	98.5	5.2	103.4	98.5	4.9	0.3
8	A	99.7	95.8	3.9	98.7	95.8	2.9	1.0
9	A	101.3	96.7	4.6	101.8	96.7	5.1	−0.5
10	A	96.5	91.6	4.9	96.8	90.2	6.6	−1.7
1	B	99.8	94.5	5.3	100.4	95.3	5.1	0.2
2	B	102.0	99.4	2.6	101.1	96.1	5.0	−2.4
3	B	98.0	90.5	7.5	97.6	91.1	6.5	1.0
4	B	98.1	92.5	5.6	97.4	92.4	5.0	0.6

(continued)

(continued)

Table 5.5 The two-time-period approach.

Part	Appraiser	Rep 1	Rep 2	Dif 1	Rep 3	Rep 4	Dif 1	Dif 2
5	B	97.5	91.7	5.8	97.2	93.3	3.9	1.9
6	B	98.5	92.8	5.7	98.1	94.0	4.1	1.6
7	B	101.0	99.4	1.6	103.5	96.3	7.2	−5.6
8	B	96.6	93.5	3.1	99.4	94.1	5.3	−1.8
9	B	100.8	94.0	6.8	98.8	95.3	3.5	3.3
10	B	95.6	90.5	5.1	96.0	89.0	7.0	−1.9
1	C	99.9	94.3	5.6	100.8	96.3	4.5	1.1
2	C	100.5	96.2	4.3	103.1	95.9	7.2	−2.9
3	C	97.9	90.2	7.7	97.1	89.8	7.3	0.4
4	C	95.7	91.5	4.2	96.2	91.4	4.8	−0.6
5	C	97.4	90.7	6.7	96.5	91.9	4.6	2.1
6	C	98.2	91.1	7.1	98.0	93.3	4.7	2.4
7	C	100.7	96.4	4.3	101.6	97.5	4.1	0.2
8	C	98.3	94.0	4.3	99.2	92.3	6.9	−2.3
9	C	98.6	93.8	4.8	100.3	95.2	5.1	−0.3
10	C	94.3	91.1	3.2	94.4	92.1	2.3	0.9

AN ANALYSIS APPROACH TO NONREPEATABLE MSAS

The design solution to the problem of contamination of the repeatability with part-to-part changes depends on a model of how the changes occur. If the delta1, delta2, and delta3 quantities in the last example did not remain constant but could change even in the stabilized conditions, the approach presented there would not work. But if there is a model that can be fitted (De Mast, 2005), it is also possible to minimize or accommodate the contamination effects through analysis alone. For example, if the assumptions that were made for the caliper model are correct, it might not be necessary to run the whole MSA twice after the stabilization period. Instead, it might be possible to model the patterns of change and remove this effect mathematically. Here is an example using the original data presented in Table 5.1. The model assumes that the change in part with repetition is a linear trend depending only on the repetition number. The analysis is accomplished by including a covariate that fits and removes the linear trend simultaneously with estimation of the measurement systems variance components. This is shown in Table 5.6.

Table 5.6 An MSA with a linear trend removed.

Error Source	Variance Estimate	Standard Deviation	6*Std Dev	100*6*SD/tol range
Parts	5.96	2.44	—	—
Appraiser	1.33	1.15	6.92	8.6%
Repeatability	0.83	0.91	5.47	6.8%
RandR	2.16	1.82	10.9	13.6%

The trend that is fitted has an intercept of 98.80 and a slope of –0.9353. The naive approach to this problem yielded an RandR of 24.8 percent, which is a 45 percent reduction by using this analytic approach. Besides this simplistic trend that depends entirely on the repetition number, there are many other trends and patterns that can be handled easily with this covariate approach. For example, the delta changes between parts because of repeats might also depend on the time taken between the repeats. Or it might be a nonlinear trend. Perhaps the pattern starts with a rapid change and then eventually decreases until the change is a constant number. In either case the trend can be fitted as the time between measurements or as a decreasing function of repeat numbers, respectively.

Chapter 5 Take-Home Pay

1. The standard MSA can overestimate error when the parts change value during the study.

2. The part evolution can often be mitigated by waiting for product stabilization.

3. The part evolution can often be mitigated by duplicating the whole MSA study.

4. Covariance analysis can remove the part evolution effect for linear trends.

6

Destructive Measurement Systems

When the measurement is truly destructive in the sense that there is no possibility of a repeat measurement, the contamination from part changes can be quite difficult to extract. But often there are options that can be pursued that are similar to those applied to the nonrepeatable systems discussed in the previous chapter.

THE NAIVE APPROACH

Just as in the nonrepeatable situation, it might be feasible to arbitrarily assign pairs or triples of repeats and run the MSA in the naive way as if these were true repeats (see Figure 6.1). If the repeatability and reproducibility errors are still small enough for their application—that is, their ratios are still less than 10 percent of the tolerance width—it is safe to assume that the true measurement repeatability is smaller than this contaminated estimate and therefore is also adequate for the application needs. The only

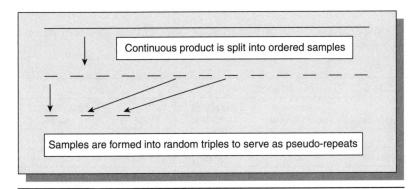

Continuous product is split into ordered samples

Samples are formed into random triples to serve as pseudo-repeats

Figure 6.1   Creating random repeats.

case wherein this assumption that the repeatability error is less than the parts-contaminated measurement repeatability is misleading is when there is a strong negative correlation between the part changes and the repeat changes. A random pairing or tripling should avoid this problem if it is worrisome. For many practical measurement systems this seems to be an issue of no importance.

Using the data supplied in Table 5.2 it is possible to analyze these in a naive way simply by creating random repeats. All measurements for appraiser A might be randomly regrouped three at a time to form the first set of pseudo-repeats. In a similar fashion, other triplets of measurements could be formed for each appraiser and treated as parts in a naive MSA. Table 6.1 shows the same data as in Table 5.2 but regrouped into these pseudo-repeats.

Table 6.1 Randomly formed pseudo-repeats.

Part	Appraiser A			Appraiser B			Appraiser C		
	Rep 1	Rep 2	Rep 3	Rep 1	Rep 2	Rep 3	Rep 1	Rep 2	Rep 3
I	99.5	100.2	99.8	89.9	98.0	92.1	91.1	92.3	98.1
II	99.8	95.6	90.0	107.0	98.8	105.3	98.7	102.0	102.3
III	105.6	96.2	99.3	100.4	94.4	103.0	106.6	100.6	96.2
IV	99.5	95.6	104.8	105.8	102.0	92.4	97.0	104.3	100.2
V	99.4	103.1	100.4	104.0	93.4	96.4	94.4	94.0	103.1
VI	104.6	105.3	103.8	93.6	95.3	101.3	103.2	90.3	94.7
VII	97.0	103.3	101.1	103.3	99.8	97.2	108.1	89.5	92.6
VIII	103.9	94.4	95.5	103.1	103.0	98.2	99.6	102.4	94.5
IX	99.6	109.7	100.3	96.6	98.5	92.8	101.2	96.4	99.3
X	96.4	95.7	104.8	98.5	93.0	98.3	96.0	96.4	96.3

The analysis results derived from the MSA of these randomly associated triplets are given in Table 6.2.

Table 6.2 The naive MSA approach to a nonrepetitive process.

Error Source	Variance Estimate	Standard Deviation	6*Std Dev	100*6*SD/tol range
Parts	0.00	0.00	—	—
Appraiser	0.58	0.76	4.57	5.7%
Repeatability	20.38	4.51	27.09	33.9%
RandR	20.96	4.58	27.47	34.3%

BLENDING APPROACH

The arbitrary association of measurements into groups that are treated as pseudo-repeats is quite common. Sometimes, rather than being accomplished directly by grouping for the MSA study, it is accomplished by preparing samples that are blended from several lots, as depicted in Figure 6.2. If the blending process is random with respect to the measurement process, any selection of pseudo-repeats will mimic the random approach (Doganaksoy, 1996). Usually the blending will inflate the estimate of repeatability but will provide fair treatment of the appraiser effects.

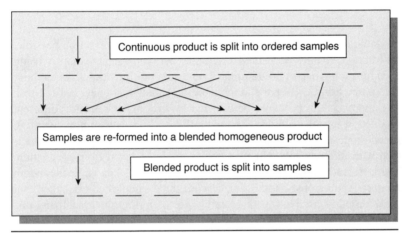

Figure 6.2 Blending to achieve repeats.

THE STANDARDIZATION APPROACH

Another way to alter the part-to-part variation so that pseudo-repeats will be almost as good as true repeats for a destructive measurement system is to homogenize or standardize the material, as shown in Figure 6.3. Complete or ideal homogenization would imply that every piece or subvolume of the material is nearly identical. For the MSA this would further imply that there were no part changes that could interfere with the repeatability estimate, no matter the choice of pseudo-repeats.

Realistically, it is not possible to make all the samples the exact same value. Additionally, this might not be the best recommended scenario in which to conduct a measurement study, because one typically wants to apply the measurement system to a variety of material values covering the range encountered in a real application. It is more reasonable, and probably more

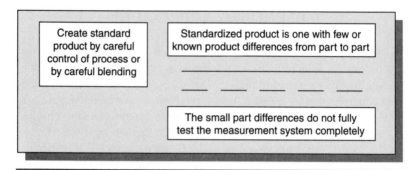

Figure 6.3 The standardization approach.

desirable as well, to have different scales of effects. It might be possible, for example, to have very small differences between the parts that are being used for each group of pseudo-repeats and larger differences between those being used as different parts in the tableau. Although not necessarily ideal, this approach can probably be effected by making or selecting different production lots of materials. Assuming that each lot is well homogenized, pseudo-repeats are chosen from a unique lot, with parts becoming associated with different lots. Ninety samples might be taken from production. Then, 10 random mixtures are made to represent the 10 parts. Each random mixture is laboriously homogenized until three samples can be taken that can be used as pseudo-repeats. See Figure 6.4 for help in visualizing this process.

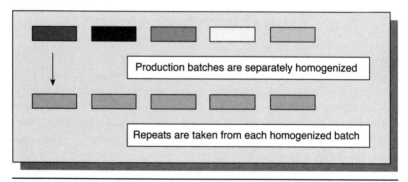

Figure 6.4 Homogenized lots used for repeats.

A MODELING APPROACH

An approach to decoupling the part changes and the repeatability differences can also be implemented through an approach that correctly models the pattern of changes. If one knows or can find the pattern by which the parts change, one can use this information to advantage in the design of the MSA. For example, there are textile cord witness reels that are collected from long continuous runs of a specific cord type. These witness lots are specially manufactured to be as homogeneous as possible, but usually there are still differences and trends between sections of the cord. These patterns can be caused by manufacturing cycles or by the method of winding or even by aging.

One can assume that there is a cyclic variation in the part values coming from the role with an unknown periodicity and phase (Schwartz, 1975). For the purpose of conducting an MSA study on a break strength test machine, the laboratory technician might select 30 samples consecutively from this reel. If these samples are randomly associated into pairs or triples, the part differences are likely to be similar in each set of pseudo-repeats, and one should arrive at a fair estimate of repeatability plus part differences similar to the approach outlined earlier. This purposely confounds the part variation with the repeatability variation in a balanced but inefficient way. There is no way to disassociate the part and repeat effects in this approach.

Another approach is that the test procedure might run the samples on the test machine in the same order in which they were pulled from the witness bobbin. This means that the cycle that is present in the witness material should be retained in the sets of repeats. If one can model or understand or adjust correctly for the pattern of part differences, it might be more advantageous to leave this pattern embedded in the repeat pattern rather than smear its effect through the formation of random pseudo-repeats.

It is unlikely that the pattern and the sample will be exactly coincident, as in the previous example, but it might still be possible to pull out all the product effects in the estimation of the repeatability effects. If the pattern is a single cosine curve that occurs every meter and each sample is 0.4 meters long, the pattern of product changes associated with the sample will look like that in Figure 6.5.

The easiest way to remove the part variation when a pattern is known or suspected is to use regression in the form of analysis of covariance (ANCOVA) (De Mast, 2005). This process can be accomplished in two sequential steps or in one more complicated analysis depending on one's available software.

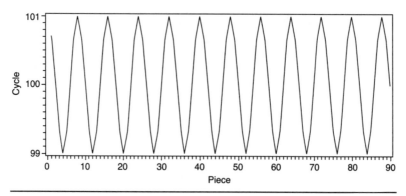

Figure 6.5 The underlying pattern in the breaking strength example.

A Two-Step Analysis Approach

In the two-step analysis one first fits the trend in product changes in step one and then performs the usual MSA on the residuals from this initial analysis (Ott, 1984). The *residuals* are the differences between the fitted trend and the measured value and, in this example, encompass both appraiser and repeatability effects. In step two, one then computes the MSA on these residuals as if they were the raw measurement values.

If one knows the exact relationship between part value and position in bobbin, there would be no part variation left in the repeatability estimate. In reality, the pattern may not be precise or one's knowledge of it may not be exact, so there will be some error left in the procedure. Still, it is feasible in this way to remove most of the part contamination from the repeatability and reproducibility estimates, and this is a great boon. Table 6.3 shows the details of the fitting of the cosine curve that is done in step one, and Table 6.4 shows the MSA results using the residuals from this first step.

A One-Step Analytic Approach

It is also possible to conduct a one-step analysis if one has access to an ANOVA program that allows continuous or covariate terms. For example, the excellent statistical package SAS can be used in this way quite easily using PROC MIXED. The results are given in Table 6.5. Notice that the results are very close in values to the two-step approach.

Table 6.3   The results of the first step of fitting.

Source	DF	Sum of Squares	Mean Square	F Value	Pr > F
Model	1	975.46592	975.46592	98.38	<.0001
Error	88	872.57863	9.91567	—	—
Corrected Total	89	1848.04456	—	—	—
Root MSE	3.14892	R-Square	0.5278	—	—
Dependent Mean	98.90778	Adj R-Sq	0.5225	—	—
Coeff Var	3.18369	—	—	—	—

Variable	DF	Parameter Estimate	Standard Error	t Value	Pr > ltl
Intercept	1	567.16786	47.21204	12.01	<.0001
Cycle	1	−4.68223	0.47207	−9.92	<.0001

Table 6.4   The MSA results on the residual from the first step.

Error Source	Variance Estimate	Standard Deviation	6*Std Dev	100*6*SD/tol range
Parts	6.40	2.53	—	—
Appraiser	1.29	1.14	6.81	8.5%
Repeatability	3.10	1.76	10.60	13.2%
RandR	4.39	2.10	10.60	15.7%

Table 6.5   The results of the one-step ANCOVA.

Error Source	Variance Estimate	Standard Deviation	6*Std Dev	100*6*SD/tol range
Parts	6.43	2.54	—	—
Appraiser	1.31	1.14	6.87	8.6%
Repeatability	3.12	1.76	10.60	13.2%
RandR	4.43	2.10	12.63	15.8%
Intercept	586.09	Slope	−4.87	—

Finding Product Change Patterns

Another advantage of covariance analysis applied to MSA studies occurs when the investigator does not have an exact idea of the pattern by which the product changes. In the last example the pattern is a periodic pattern. If one suspected that there was a periodic pattern but did not know the frequency or magnitude, one could utilize covariance analysis to model different possibilities and choose a model that fits well. For example, one might construct terms of different frequencies and put all of them into the covariance analysis. Under the right conditions the coefficients for the true effect will stand out significantly and only their removal will drastically affect the residuals and hence the estimate of measurement errors. The statistical ignorance and lack-of-fit diagnostics that come standard with the SAS package can help one certify what these effects are.

Here is an example using the same data as the last example, in which there is a single cyclic pattern to the product changes in the breaking strength of a textile cord. In this case the analyst, without any clear knowledge of the precise frequency, only suspects that there is a cycle. The analyst creates terms that represent cycles that run 1–10 times through the data and depends on the regressions significance testing to identify the correct term. Looking at the significance of each term displayed in Table 6.6, it is clear that Cycle8 is dominant in its effects. This term is identical to the one assumed to be correct in the last example. It is possible to rerun the regression using just the single identified term, or it is also possible to leave the regression alone with its multiple cycles and count on the fact that unimportant terms will tend to have small coefficients and hence will change the results by a small amount.

Table 6.6   The results of fitting terms representing eight different cycles.

Variable	DF	Parameter Estimate	Standard Error	*t* Value	Pr > |*t*|
Intercept	1	98.94413	0.33800	292.73	<.0001
Cycle2	1	−0.11426	0.33796	−0.34	0.7362
Cycle3	1	−0.94520	0.47805	−1.98	0.0515
Cycle4	1	−0.06836	0.47878	−0.14	0.8868
Cycle5	1	−0.02609	0.47827	−0.05	0.9566
Cycle6	1	−0.20939	0.47912	−0.44	0.6633
Cycle7	1	0.02507	0.48342	0.05	0.9588
Cycle8	1	−4.68853	0.48712	−9.62	<.0001
Cycle9	1	0.03752	0.48216	0.08	0.9382
Cycle10	1	−0.39601	0.47940	−0.83	0.4112

As an example, Table 6.7 details the results of an MSA on the residuals from the regression analysis with all eight cyclic terms. Again, the results are very similar to other methods for removal of these effects.

Cyclic patterns in data are often studied through the techniques of Fourier analysis. In this kind of analysis the signal is split into independent components, each of which is a cosine or sine curve with a different frequency. The details of the computations ensure that there are curves that pass an integral number of times through the full signal—once, twice, and so on. The computations also ensure that the components sum to the full signal. Fourier analysis can be performed with regression analysis by constructing the appropriate independent inputs. Each input must represent the same cosine and sine curves that are needed in the full decomposition.

There are other ways in which this covariance approach can be extended in the analysis of destructive measurement systems. For example, the pattern might be more complex, perhaps changing frequency in the middle of the test. Or there might be a two-dimensional pattern (Phillips, 1997). Often rubber is produced in sheets that can be sampled lengthwise and crosswise. There are likely to be patterns in both directions that affect the true product value. The covariance analysis can be modified to accommodate this two-dimensional pattern quite readily. Figure 6.6 shows an example of a

Table 6.7 The MSA results on the residual from the eight-cycle fit.

Error Source	Variance Estimate	Standard Deviation	6*Std Dev	100*6*SD/tol range
Parts	6.39	2.53	—	—
Appraiser	0.64	0.90	4.90	6.0%
Repeatability	2.99	1.73	10.4	13.0%
RandR	3.63	1.91	11.4	14.3%

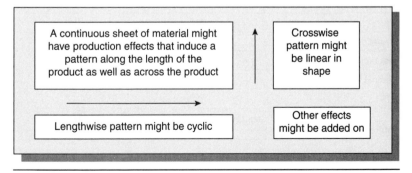

Figure 6.6 Example of a two-dimensional pattern in product.

two-dimensional pattern. A two-dimensional pattern can be coded into the covariate analysis as two variables, one representing the location along the length and one representing the location along the width. A third variable could be added to represent the location of the third dimension of depth if the samples were extracted from larger blocks of material. Table 6.8 shows example data with a two-dimensional pattern, and Table 6.9 illustrates the analysis results.

Table 6.8   Data from a destructive test with a two-dimensional pattern.

Part	Appraiser A			Appraiser B			Appraiser C		
	Left	Center	Right	Left	Center	Right	Left	Center	Right
I	105.3	97.6	97.8	103.1	94.4	100.0	99.3	96.4	101.1
II	101.1	104.8	99.0	102.0	102.0	100.8	104.6	96.2	103.2
III	97.6	99.8	97.6	96.5	93.0	100.0	98.6	90.3	106.3
IV	97.5	94.4	101.6	94.4	92.1	103.3	92.7	92.3	105.2
V	98.4	95.5	105.3	96.5	94.4	107.8	92.0	97.0	105.1
VI	97.4	96.4	107.6	95.2	98.2	105.0	92.5	102.3	98.4
VII	101.9	104.6	111.7	98.4	107.0	105.0	97.6	108.1	98.4
VIII	97.3	103.8	102.2	94.6	103.3	95.4	94.3	102.4	91.5
IX	98.3	104.8	97.7	96.3	105.3	94.8	98.2	98.7	94.6
X	94.2	99.5	92.9	93.3	93.6	91.9	96.1	91.1	98.0

Table 6.9   The MSA results from the two-dimensional pattern.

Error Source	Variance Estimate	Standard Deviation	6*Std Dev	100*6*SD/tol range
Parts	6.43	2.54	—	—
Appraiser	1.31	1.14	6.87	8.6%
Repeatability	3.16	1.78	10.70	26.7%
RandR	4.45	2.11	12.70	15.8%

Intercept	98.95	Lengthwise Slope	−4.88	Crosswise Slope	2.05

Chapter 6 Take-Home Pay

1. Destructive measurements also misinterpret part change as error.

2. This part misinterpretation can be ignored if it is small.

3. This part misinterpretation can be mitigated by blending.

4. Part misinterpretation can be mitigated by the use of homogenized material.

5. Covariance analysis can remove the part change effect for periodic effects.

7

In-Line and Dynamic Measurement Systems

SOME MEASUREMENT SYSTEMS WITHOUT TRUE REPEATS

There are two other situations that are exceedingly common in industrial and manufacturing situations and to which it is difficult to apply the standard MSA: when the measurement processes are dynamic and when they cannot be taken off-line. A classic example is a process in which there is an in-line weight scale. This scale might be situated near the end of an industrial dry-goods process. A bulk material such as flour or cereal is transported along a conveyor belt system leading to a hopper with an in-line scale. The material accumulates on this scale until the target weight is reached, and then the product is released into a package. The scale is reset, and the next lot of material begins accumulating. This process is repeated over and over again as new packages are filled. This is illustrated in Figure 7.1. If the scale underestimates the weight, customers could feel cheated; an overestimate in

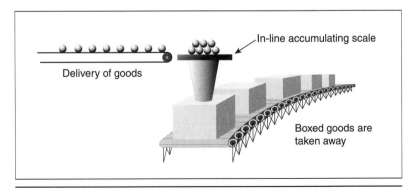

Figure 7.1 An in-line weight scale.

weight might lead to losses in profitability. It is critical to maintain a small measurement error on this scale, and yet it is difficult to perform a standard MSA. This is because any particular product is seen once, and it is difficult to create a true repeat. One might, for example, attempt to capture the material and reintroduce it into a conveyor system, but more than likely some loss of material will occur and the true weight will change.

Another example of an in-line measurement might be a flow meter hidden inside a transfer pipeline embedded somewhere in the complexity of a chemical processing plant. When in operation, this flow meter might sample the flow in that section of pipeline every second and transmit a value to a centralized control system, where it is recorded and used for process adjustment and monitoring. Consider Figure 7.2, which illustrates this process. If this flow meter malfunctions in such a way that it gives erroneous readings, control of the plant can be adversely affected. And yet it is nearly impossible to evaluate the measurement system properties of such a device under true process conditions because the measurement is hard to access and highly dynamic.

For these kinds of systems, which are quite common today and are becoming even more popular as rapid advancements are made in electronics and computer applications, the standard measurement systems study (such as that exampled in the AIAG guide *Measurement Systems Analysis,* third edition) will not correctly evaluate the true measurement performance of the device. This is because there is no chance of exactly repeating the same sample of material in separate trials. At best, there is some small part of the product change that is confused with measurement error, and at worst, there is almost complete distortion of the values. That is, the dynamic measurement system can look like anything from slightly deformative to completely destructive, depending on the circumstances.

This dynamic or in-line situation has a lot in common with the deformative and destructive measurement systems that were covered in chapters

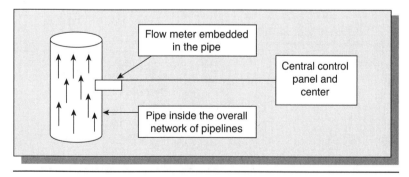

Figure 7.2   The in-process flow meter.

5 and 6. It will be seen that some of the same approaches detailed in chapters 5 and 6 can be applied to better separate the product changes from the measurement error. This chapter will provide detailed examples of the adaptation of these MSA methods for dynamic and in-line situations.

Ignore the Product Change Effects

It is always possible to just go ahead and run the standard MSA with no consideration of the potential contamination of measurement effects with product changes. In cases where the tolerances are wide enough or the product stable enough, this approach might be adequate. Essentially, the study consists of stabilizing the system and then taking consecutive measurements that can be used as pseudo-repeats. For example, consider Figure 7.3, which shows the results of taking 100 consecutive measurements on the in-line weight scale. In the naive approach one can simply extract 90 of these 100 measurements and divide them into 30 sets of three in an attempt to mimic the parts and repetitions that are required for the standard analysis. Notice that nothing is used to justify this grouping except, perhaps, the expectation that the difference in measurements will be larger when the samples are further separated in time. Whatever changes there are from measurement to measurement simply add into the estimate of repeatability. This type of arbitrary grouping is shown in Table 7.1.

One major difference among deformative, destructive, and dynamic measurements is that operator or inspector effects are often not important in dynamic situations, whereas they tend to be dominant for the other two categories. Therefore, the reproducibility is often ignored for dynamic MSA. However, one may create pseudo-operators just as the pseudo-repeats and pseudo-parts were created—that is, by simply labeling the first 30

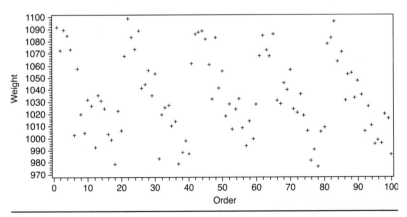

Figure 7.3 The 90 consecutive in-line scale measurements.

Table 7.1   The naive grouping of the in-line scale weights.

Part	Operator 1			Operator 2			Operator 3		
	Rep 1	Rep 2	Rep 3	Rep 1	Rep 2	Rep 3	Rep 1	Rep 2	Rep 3
1	1091	1074	1089	1084	1073	1002	1057	1019	1004
2	1031	1026	992	1035	1030	1024	1003	998	978
3	1022	1006	1067	1098	1083	1073	1081	1041	1046
4	1056	1035	1053	982	1091	1025	1026	1009	9013
5	978	988	977	986	1061	1085	1087	1088	1081
6	1060	1032	1082	1041	1055	1017	1028	1007	1024
7	1032	1008	993	1014	998	1028	1068	1084	1072
8	1067	1085	1031	1028	1045	1030	1056	1024	1020
9	1036	1018	1006	981	990	976	1005	1008	1077
10	1082	1096	1063	1071	1030	1053	1053	1043	1047

measurements as coming from operator 1, the second 30 from operator 2, and the final 30 from operator 3. This grouping is also shown in Table 7.1. Because many processes will wander as the sampling interval gets longer, it is quite possible that the pseudo-operators created in this way will show effects. These effects are, of course, not due to real operator effects in these situations but may be due to very real instabilities or slow trends that exist in the process. They may be of interest in their own right, but clearly their interpretation as measurement effects is questionable.

In this naive approach it is almost certain that product changes will contaminate the estimates of the various measurement systems study effects, such as the repeatability, reproducibility, and part variations. Although it is possible to actually decrease the variation if the product effects compensate for the measurement errors, as shown in Figure 7.3, in the vast majority of real applications the product contamination should make the apparent measurement variation look larger than it really is. Therefore, if the numbers computed on this contaminated system still indicate that the system is adequate for its application tolerances, one has good reason to believe that the actual system error is smaller and is also adequate. In Figure 7.4 the triangles represent the part values, and the cloverleafs represent the measurement effects, both of which perfectly compensate to yield the flat line representing the possible observed values.

Of course, if the results from this naive analysis do not pass muster, one does not know whether the real measurement system is sufficient or not. One must do some additional analysis or make some additional assumptions to go further. If the dynamic nature of the measurement system is not

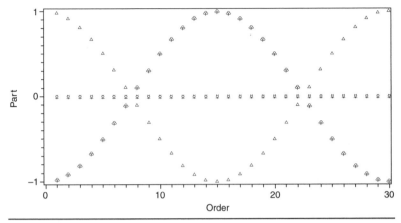

Figure 7.4   An example of possible compensation between material and measurement.

taken into account in the planning stage of the study, it might be impossible to do any kind of analysis that could separate the product and measurement changes. Most of the time when faced with this kind of situation, the investigator has no recourse but to run a new study. So it is advisable and more cost effective to plan one's study to support a successful MSA in the event that the naive approach is not sufficient. Some of these approaches are discussed later in the chapter.

Consider the results of the naive MSA computed using the ANOVA methods based on the previous data and presented here as Table 7.2. The foundational data used in this analysis are the consecutive weights from the in-line scale and have been arbitrarily separated into pseudo-parts, pseudo-operators, and pseudo-repeats as described earlier. Notice that the results do not meet the usual 30 percent RandR ratio when compared to the tolerances of +/–100 grams for this application.

Table 7.2   MSA results of naive in-line scale.

Error Source	Variance Estimate	Standard Deviation	Percent Tolerance
Parts	195.16	13.97	42
Appraisers	0 (was negative)	0	0
Repeatability	947.76	30.79	92.3
RandR	947.76	30.79	92.3

Complete Randomization of Pattern

Another variation on the theme of applying a naive MSA to a dynamic situation is the attempt to create pseudo-repeats from the raw consecutive data with the ordering destroyed. That is, the 90 consecutive observations are reselected in a completely randomized order and then grouped into pseudo-repeats, pseudo-parts, and pseudo-operators. A little reflection will convince one that this randomization should eliminate any spurious effects in the part and reproducibility estimates due to the timing of the samples. That is, the product variation that contaminated all three effects in patterned ways under the initial naive analysis when it was done in consecutive order should now contaminate all three in equal portion under complete randomization. It should be just as likely that one will see large process shifts within a set of three pseudo-repeats as between the pseudo-operators or pseudo-parts. One would expect that the MSA would show a huge repeatability problem and no other effects under this approach. Figure 7.5 shows the logic of this complete randomization, Table 7.3 details the data, and Table 7.4 shows the results of the analysis. This approach is perhaps easier to interpret because there are no doubts as to where the contamination effects are hidden, but it can still lead to rejection of adequate measurement systems simply because the analysis cannot separate out the product change contamination.

Use of Standards

Just as in the case of deformative and destructive measurement systems, it is tempting to introduce some kind of standard material into the process and

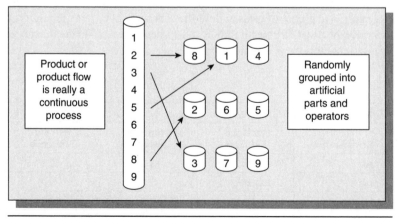

Figure 7.5   Logic of the complete randomization of in-line scale weights.

Table 7.3   The randomly mixed in-line scale data.

Part	Operator 1			Operator 2			Operator 3		
	Rep 1	Rep 2	Rep 3	Rep 1	Rep 2	Rep 3	Rep 1	Rep 2	Rep3
1	1085	996	1028	1006	1002	1045	1061	1031	1072
2	990	1032	1053	1056	1044	1026	1084	1031	999
3	1007	992	1071	1020	1035	1072	1098	1019	1089
4	1063	1032	1024	1088	1056	1006	995	1082	1019
5	982	1067	1084	1096	1026	1008	997	993	1055
6	1081	1006	1087	1028	1057	1004	1033	1036	1047
7	998	978	1013	1003	1018	978	1040	986	1035
8	1035	1016	1009	1005	1010	1017	1024	1019	1030
9	1025	986	1036	1077	1014	981	1026	1053	1060
10	1028	1073	976	1030	1088	1067	988	1073	1053

Table 7.4   The RandR results from the completely randomized in-line scale weights.

Error Source	Variance Estimate	Standard Deviation	Percent Tolerance
Parts	10.24	3.2	9.60
Appraisers	0 (was negative)	0	0
Repeatability	1078.7	32.8	98.4
RandR	1078.0	32.8	98.4

read this dynamically with the measurement system. For example, with the in-line weight scale, this might be the placement of calibrated and traceable masses on the scale in a repeated fashion. These traceable masses will certainly be of known and repeatable true weight, but does the system treat them the same as the products that it normally handles? Standard masses are often cylindrical in shape and may have a shaped head for hanging. The bulk material normally processed in the operation may be grainy and clumped. It may also be soft and break into smaller pieces during the treatment, or it could slide if it is deposited onto the conveyor in a heap. Air pockets or trapped water that the final product may not have upon packaging could also be present. In some systems it might not even be possible to introduce the standard masses into the system, as they may lodge in machinery parts or destroy sieves or filters.

It is unlikely that the system will treat the standard masses in exactly the same way as it would the actual bulk product that is normally weighted by this scale. This error could be in either direction; that is, the standard masses could be easier to measure or the bulk material could be easier to measure. The only sure assumption about the relationship between the two seems to be that at least with standard masses, the product's true weight should not change and hence contaminate the measurement error estimates. This is usually a safe assumption, but remember that the standard masses are being reintroduced into the system several times, and contamination of the masses by grease or lubricants might actually change even these highly repeatable weights. The behavior of the system must be understood to verify that the standard masses should be more easily measured than the real product. If this case proves to be reasonable, one is faced with a situation that is in some ways the opposite of the naive MSA described earlier. In the case of standards, one can be pretty sure that a failed MSA means that the true measurement systems would also fail. However, if the MSA based on standardized material succeeds, one really does not know what the situation would be for the real system. One can assume that the performance would probably be worse, but it is difficult to assess how much worse it might be. Again, this might lead to a new study. Examine Figure 7.6, which shows the process with standardized masses presented to the in-line scale. Table 7.5 shows the data from this study, and Table 7.6 shows the results.

Analytic Approach—Covariate

Modeling the Pattern

Two options will now be presented that should provide a better estimate of true measurement error than that achieved by the simple naive analysis of a dynamic measurement system. The first approach rests on the possibility of modeling the part variation as some simple function independent of the measurement effects. This model can then be used to remove (to the extent that the model is adequate) the product contamination from the measurement effects. The success of this approach relies on the proper evaluation of the model, which is meant to account for the product pattern. This pattern may not exist or it might be too complex to model adequately, but often there are simple effects in real processes that can be adequately described by manageable choices.

A plot of the 90 consecutive weights from the in-line scale is given in Figure 7.3. Notice that there appears to be an underlying trend to these weights. Remember that these weights are assumed to be consecutive—that is, close in time—and to have come from a continuous production lot. The pattern in question stems primarily from this underlying production pattern.

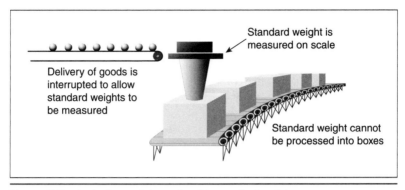

Figure 7.6 An in-line weight scale with standard weights.

Table 7.5 Standard weights used on the in-line scale.

	Operator 1			Operator 2			Operator 3		
Part	Rep 1	Rep 2	Rep 3	Rep 1	Rep 2	Rep 3	Rep 1	Rep 2	Rep 3
1	1033	998	1030	1006	986	1029	1009	1015	1020
2	990	1016	1037	1040	1028	1028	1032	1015	999
3	1009	994	1019	1022	1019	1020	1046	1003	1037
4	1011	1034	1026	1036	1040	1008	995	1066	1021
5	984	1015	1032	1044	1028	1008	997	993	1039
6	1029	1008	1035	1028	1041	988	1017	986	1035
7	1035	1016	1009	1005	1041	988	1017	1038	1031
8	1025	986	1036	1077	1014	981	1026	1053	1060
9	1028	1073	976	1030	1088	1067	988	1073	1053
10	1012	1021	976	1032	1036	1015	988	1021	1037

Table 7.6 The RandR results from the completely randomized in-line scale weights.

Error Source	Variance Estimate	Standard Deviation	Percent Tolerance
Parts	6.4	2.5	3.0
Appraisers	0 (was negative)	0	0
Repeatability	340.17	18.4	55.2
RandR	340.17	18.4	55.2

Knowledge of this production process may lead directly to a guess as to the shape of the pattern. For example, a closed-loop control system on the production line may lead to a sinusoidal pattern in the weights due to slight error in the tuning of the system or lagged time effects. In this case the pattern seems to be linear in five consecutive segments of the data. Ideally, one would research the process that created this data and investigate the likelihood of a linear effect that gets reset periodically. It is possible to simply posit this pattern, fit it, and remove it based on the assumption that it is product related and not measurement related. This assumption seems reasonable because it would be hard to imagine how the scale could be causing this kind of pattern. If there is any doubt, one should be conservative and not apply the correction. Figure 7.7 shows the data overlaid with fitted linear trends in each of the five segments.

This linear trend has an *r*-squared of 73 percent and appears to fit quite well considering the close agreement shown in the graph. If these trends can be attributed to product variation, then it can be removed before the MSA is performed. This can be done in one step if the software allows it, but a two-step process is feasible as well. This is done by computing the residuals from the fit of the linear trend and then using these in the standard MSA. Again, parts and operators are really just placeholders in this analysis. One term represents the cosine term from the Fourier analysis for harmonic 3, and one represents the sine term for the same harmonic. Table 7.7 shows the values of these residuals, and Table 7.8 shows the analysis results. Notice that these results are similar to the study run on the standard, and yet this study is done on the real process in real time.

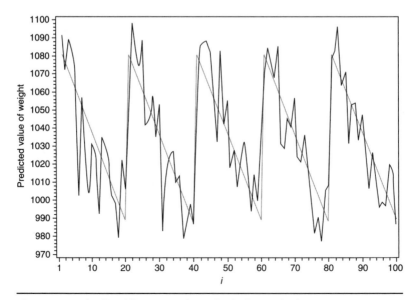

Figure 7.7 The fitted linear trends on the in-line scale data.

Table 7.7 The residuals from the in-line scale after the removal of the linear trend.

Part	Operator 1			Operator 2			Operator 3		
	Rep 1	Rep 2	Rep 3	Rep 1	Rep 2	Rep 3	Rep 1	Rep 2	Rep 3
1	10.1	−4.1	17.7	17.7	11.4	−54.1	5.3	−27.4	−38.1
2	−5.6	−6.4	−35.2	12.7	12.9	11.5	−4.9	−4.8	−19.9
3	28.7	17.7	−13.5	22.0	11.7	7.1	26.9	−15.9	−7.8
4	8.8	−7.2	15.8	−49.8	−8.1	2.1	8.9	−3.2	5.4
5	−24.4	−10.2	4.3	−1.8	−19.7	9.4	15.7	21.8	19.9
6	3.9	−19.5	35.6	−0.8	18.0	−14.6	0.6	−15.4	6.3
7	19.4	0.3	−9.7	16.1	5.6	39.3	−13.0	8.2	1.0
8	1.2	24.1	−25.8	−23.1	−1.7	−2.2	19.0	−8.3	−6.8
9	13.3	0.6	−7.0	−26.9	−12.5	−21.6	11.7	20.0	−3.3
10	6.5	24.7	−3.4	9.2	−25.9	0	6.5	−9.2	9.9

Table 7.8 The RandR results from the trend corrected in-line scale weights.

Error Source	Variance Estimate	Standard Deviation	Percent Tolerance
Parts	0 (was negative)	0	0
Appraisers	0.11	0.33	1.0
Repeatability	321.11	17.9	53.8
RandR	321.22	17.9	53.8

Analytic Method—Matching

If there is a consistent pattern in the product weights that can be identified through Fourier analysis or some other data exploration technique, then there is an additional way to reduce or remove the product change contamination of the measurement system error components. This second method is accomplished by matching the observations in such a way that they most closely resemble each other in terms of where they occur in the pattern. If one again examines the plot of the in-line scale measurements, one can imagine that if the first observations from each segment are combined they will be at roughly equivalent areas of the underlying pattern. Note that this matching must be done on the underlying pattern and not directly on the observations themselves. If the matching is done on the measurements, it is quite likely that the measurement error will be artificially reduced along with the product variations.

Specifically, for the in-line scale data, one matched group is formed of observations 1, 21, 41, 61, and 81. Another grouping is observations 2, 22, 42, 62, and 82, with other groups being formed along the same pattern. If one considers these groupings in light of the linear trend lines, they are formed from points taken at the same section of the lines along these trends. Because the observations do not break up into groups of three, as would best fit into the standard MSA format, it is probably better in this case to consider each set of matched observations as a distinct part with only one operator. In this way, 10 "parts" will have five repeats and 10 "parts" will have only four repeats. Consider Table 7.9, which shows the setup of this matching approach. The MSA is then run on this data, ignoring operator effects, to arrive at the results shown in Table 7.10.

Table 7.9 The matched observations for the in-line scale analysis.

Part	Repeat 1	Repeat 2	Repeat 3	Repeat 4	Repeat 5
1	1091	1067	1061	1068	1077
2	1072	1098	1085	1084	1082
3	1089	1083	1087	1072	1096
4	1084	1073	1088	1067	1063
5	1073	1088	1081	1085	1071
6	1003	1041	1060	1031	1030
7	1057	1044	1032	1028	1053
8	1019	1056	1082	1045	1053
9	1004	1035	1041	1040	1033
10	1031	1053	1055	1056	1047
11	1026	982	1017	1024	—
12	992	1019	1028	1020	—
13	1035	1025	1007	1036	—
14	1030	1026	1024	1018	—
15	1024	1009	1032	1006	—
16	1003	1013	1008	981	—
17	998	978	993	990	—
18	978	988	1014	976	—
19	1022	997	999	1005	—
20	1006	986	1028	1008	—

Table 7.10 The RandR results of the matched study on the in-line scale.

Error Source	Variance Estimate	Standard Deviation	Percent Tolerance
Parts	952.63	30.86	92.6
Appraisers	0 (only one operator)	0	0
Repeatability	197.41	14.0	42.1
RandR	197.41	14.0	42.1

Chapter 7 Take-Home Pay

1. Dynamic systems do not allow remeasurement of the same part.

2. Small part changes can be ignored.

3. Biases can be eliminated by complete randomization of the time order.

4. Sometimes standard parts can be used to mitigate the effect.

5. Covariance analysis can be used to separate the part change and the error.

6. Matching of sample points can be used to mitigate the effect.

8

The Design of MSA Studies

SOME BASIC DESIGN NEEDS

The proper analysis of a measurement systems study requires that the study be set up correctly and executed properly—that is, in a fashion that enables the intended analysis. A study that seeks to isolate only the effect of repeatability can afford to be much simpler than one that seeks to estimate reproducibility as well. A study that seeks to take advantage of the filtering properties of multiple factors in an MSA will need a more complex design. It is critical that the design allow the investigator to cleanly and clearly separate the different sources of measurement variation. One objective of this chapter is to establish ground rules to guarantee this clean estimation of the effects.

Another feature of an MSA study that is critical is the number of elements used in the study. As was demonstrated in Chapter 3, too few samples will make the variation in the estimates so large as to make the results meaningless. On the other hand, collecting more samples than is necessary is a waste of resources. The requirements become even more complicated when there are different requirements for precision for each of the multiple sources of variation. For example, a study that wants equal variation of estimates for both parts and appraisers requires a different design than the one provided by the standard MSA. Another objective of this chapter is to explain approaches to answering this sample-size question.

An objective of a good study is that the study allow the possibility of discovering new facts about the system. Although planning for the unexpected is difficult, it is not impossible. For example, one can plan to collect background information that may be needed in case of problems in the analysis. And one can also conduct the study in ways that may allow the segregation or separation of the data after the initial analysis. A third objective of this chapter is to explain how this planning for the unexpected may be accomplished.

THE ADEQUATE ISOLATION OF EFFECTS

A Full Factorial Approach

One type of design that allows the separation of the effects required for a good MSA is the standard full factorial, or crossed, design (Raktoe, 1981). This kind of design is suitable for discrete settings and consists of all combinations of the settings of each source. Thus, all possible combinations of settings occur at least once in the study. A study with four sources, each with three settings, would require $3 \times 3 \times 3 \times 3 = 81$ combinations under this plan. The standard MSA study is this kind of design. It has three sources—called parts, appraisers, and repetitions—each with 10, 3, and 3 levels, respectively. The total number of combinations required for this full factorial is therefore $10 \times 3 \times 3 = 90$ measurements. As will be shown, this design is actually a little inefficient to achieve the purposes outlined for the standard MSA.

As an example of this full factorial approach, consider the example given earlier in which there are four sources of variation to be estimated: parts, appraisers, setups, and repeatability. Each of these sources had three settings that will be used in the study. All together, this means that there are 81 combinations that must be performed. See Table 8.1 for an example of such a design. Notice that this is not a good choice for sample size, but it is adequate for the purpose of separating effects. Table 8.2 shows the analysis.

Table 8.1   A full factorial MSA design.

Obs	Seq	Part	App	Setup	Meas
1	1	1	1	1	1004.80
2	1	1	1	1	1005.32
3	1	1	1	1	1005.03
4	1	1	1	2	1002.19
5	1	1	1	2	1002.17
6	1	1	1	2	1002.16
7	1	1	1	3	1002.44
8	1	1	1	3	1002.80
9	1	1	1	3	1002.47
10	1	1	2	1	1003.51
11	1	1	2	1	1003.62
12	1	1	2	1	1003.48
13	1	1	2	2	1000.72

(continued)

Table 8.1 A full factorial MSA design.

Obs	Seq	Part	App	Setup	Meas
14	1	1	2	2	1001.08
15	1	1	2	2	1001.12
16	1	1	2	3	1001.10
17	1	1	2	3	1000.99
18	1	1	2	3	1001.24
19	1	1	3	1	1003.91
20	1	1	3	1	1003.88
21	1	1	3	1	1003.07
22	1	1	3	2	1001.29
23	1	1	3	2	1001.07
24	1	1	3	2	1000.75
25	1	1	3	3	1001.34
26	1	1	3	3	1001.69
27	1	1	3	3	1001.44
28	1	2	1	1	1003.53
29	1	2	1	1	1003.67
30	1	2	1	1	1003.61
31	1	2	1	2	1001.29
32	1	2	1	2	1000.98
33	1	2	1	2	1001.40
34	1	2	1	3	1001.42
35	1	2	1	3	1001.34
36	1	2	1	3	1001.49
37	1	2	2	1	1002.37
38	1	2	2	1	1002.41
39	1	2	2	1	1002.39
40	1	2	2	2	999.38
41	1	2	2	2	999.32
42	1	2	2	2	999.13
43	1	2	2	3	999.32
44	1	2	2	3	999.81
45	1	2	2	3	999.71
46	1	2	3	1	1001.89
47	1	2	3	1	1001.73

(continued)

(continued)

Table 8.1   A full factorial MSA design.

Obs	Seq	Part	App	Setup	Meas
48	1	2	3	1	1002.31
49	1	2	3	2	999.68
50	1	2	3	2	999.24
51	1	2	3	2	999.69
52	1	2	3	3	999.68
53	1	2	3	3	1000.33
54	1	2	3	3	999.72
55	1	3	1	1	998.82
56	1	3	1	1	998.54
57	1	3	1	1	998.70
58	1	3	1	2	996.06
59	1	3	1	2	995.99
60	1	3	1	2	995.97
61	1	3	1	3	996.60
62	1	3	1	3	996.45
63	1	3	1	3	996.55
64	1	3	2	1	997.59
65	1	3	2	1	997.74
66	1	3	2	1	997.70
67	1	3	2	2	994.79
68	1	3	2	2	994.72
69	1	3	2	2	994.22
70	1	3	2	3	995.08
71	1	3	2	3	995.02
72	1	3	2	3	994.91
73	1	3	3	1	997.46
74	1	3	3	1	997.02
75	1	3	3	1	997.21
76	1	3	3	2	994.14
77	1	3	3	2	994.81
78	1	3	3	2	994.54
79	1	3	3	3	995.13
80	1	3	3	3	995.09
81	1	3	3	3	994.87

Table 8.2   Full factorial design results.

Error Source	Variance Estimate	Standard Deviation	Percent Tolerance
Parts	10.78	3.28	—
Appraiser	0.69	0.83	12.4
Setup	2.17	1.47	22.0
Repeatability	0.06	0.24	3.6
Uncertainty	2.91	1.71	25.6

Another Way of Treating the Repeatability in the Full Factorial

The standard MSA and factorial designs similar to it can be reduced somewhat because of the special role of repeatability. If the residual component of the ANOVA is used as the estimate of repeatability, one does not need to treat it in the same way as the other crossed factors. It is apparent from the ANOVA table that most of the study information goes to the estimation of this residual. An MSA design that has a single repeat for each combination will still provide an estimate of this residual error and hence of repeatability. So in some sense, two-thirds of the study is given over to improving this estimate of repeatability. Understanding this fact leads to another compromise in which some combinations have true repeats and others have only a single measurement. As an example of this staggered approach, consider the example data given earlier but altered to demonstrate this new approach. In this altered dataset, shown in Table 8.3, only one-third of the combinations

Table 8.3   The unequally replicated MSA.

Obs	Seq	Part	App	Setup	Rep	Meas
1	1	1	1	1	1	1004.80
2	1	1	1	2	1	1002.19
3	1	1	1	3	1	1002.44
4	1	1	2	1	1	1003.51
5	1	1	2	2	1	1000.72
6	1	1	2	3	1	1001.10
7	1	1	3	1	1	1003.91
8	1	1	3	2	1	1001.29
9	1	1	3	3	1	1001.34
10	1	2	1	1	1	1003.53

(continued)

(continued)

Table 8.3 The unequally replicated MSA.

Obs	Seq	Part	App	Setup	Rep	Meas
11	1	2	1	2	1	1001.29
12	1	2	1	3	1	1001.42
13	1	2	2	1	1	1002.37
14	1	2	2	2	1	999.38
15	1	2	2	3	1	999.32
16	1	2	3	1	1	1001.89
17	1	2	3	2	1	999.68
18	1	2	3	3	1	999.68
19	1	3	1	1	1	998.82
20	1	3	1	1	3	998.70
21	1	3	1	2	1	996.06
22	1	3	1	2	2	995.99
23	1	3	1	2	3	995.97
24	1	3	1	3	1	996.60
25	1	3	1	3	2	996.45
26	1	3	1	3	3	996.55
27	1	3	2	1	1	997.59
28	1	3	2	1	2	997.74
29	1	3	2	1	3	997.70
30	1	3	2	2	1	994.79
31	1	3	2	2	2	994.72
32	1	3	2	2	3	994.22
33	1	3	2	3	1	995.08
34	1	3	2	3	2	995.02
35	1	3	2	3	3	994.91
36	1	3	3	1	1	997.46
37	1	3	3	1	2	997.02
38	1	3	3	1	3	997.21
39	1	3	3	2	1	994.14
40	1	3	3	2	2	994.81
41	1	3	3	2	3	994.54
42	1	3	3	3	1	995.13
43	1	3	3	3	2	995.09
44	1	3	3	3	3	994.87

Table 8.4   MSA analysis of the unequally replicated design.

Error Source	Variance Estimate	Standard Deviation	Percent Tolerance
Parts	13.78	3.71	—
Appraiser	0.55	0.74	11.1
Setup	2.19	1.48	22.2
Repeatability	0.06	0.24	3.64
Uncertainty	2.80	1.67	25.0

keep their full set of repeats. Table 8.4 shows the analysis of this altered dataset, which can be compared to the previous analysis to see that there is no great direct impact from this modification.

A Fractional Factorial Approach

Although the full factorial is an easy way to ensure that the design of the MSA is adequate, for many purposes it is overkill. By choosing appropriate subsets of the full factorial, it is still possible to separate the error sources in a valid manner. One such approach is through the fractional factorial approach. A *fractional factorial* is a specifically chosen subset of the full set of combinations that still allows the ANOVA procedure to separate independent components of variation to the measurement error sources. The selection of a random subset or an improperly chosen subset will usually end in an analysis that does an inadequate job of separation and can lead to results that are less valid than those of either the full factorial or the properly chosen fractional factorial (Ryan, 1989). Keep in mind that fractional factorials are not the only way in which smaller designs can be generated for a valid MSA study, but they are probably the most common.

Consider again the example presented as Table 8.1, which is a full factorial (all combinations) design for four sources, each with three levels or representatives. These four sources are parts, appraisers, setups, and repeatability. Note that the three elements that are actually used to represent these effects are assumed to be random choices from the ranges present in the operational measurement system. Specifically, they are assumed not to be specially chosen to represent the worst or best conditions of these sources. The full factorial study requires $3 \times 3 \times 3 \times 3 = 81$ measurements to be taken. But if no interactions are thought important and one does not need to isolate them in the ANOVA, then it is possible to get the same lines in the ANOVA table with much fewer trials. The 3^4 orthogonal array looks like Table 8.5 in a generic format in which the three levels are represented by the numbers 1, 2, and 3. This array has only 18 runs rather than 81. The analysis of this data is given in Table 8.6.

Table 8.5   The fractional factorial MSA design.

Obs	Seq	Part	App	Setup	Meas
1	1	1	1	1	1004.80
2	1	1	1	1	1005.32
3	1	1	2	2	1000.72
4	1	1	2	2	1001.08
5	1	1	3	3	1001.34
6	1	1	3	3	1001.69
7	1	2	1	2	1001.29
8	1	2	1	2	1000.98
9	1	2	2	3	999.32
10	1	2	2	3	999.81
11	1	2	3	1	1001.89
12	1	2	3	1	1001.73
13	1	3	1	3	996.60
14	1	3	1	3	996.45
15	1	3	2	1	997.59
16	1	3	2	1	997.74
17	1	3	3	2	994.14
18	1	3	3	2	994.81

Table 8.6   The analysis of the fractional factorial MSA.

Error Source	Variance Estimate	Standard Deviation	Percent Tolerance
Parts	10.54	3.25	—
Appraiser	0.82	0.91	13.6
Setup	2.08	1.44	21.6
Repeatability	0.12	0.35	5.2
Uncertainty	3.02	1.74	26.1

Using Only a Single Repeat

The use of a fractional factorial design is a tremendous efficiency in generating a valid separation of the measurement dispersion sources. Consider again the reduction from 81 required runs to just 18 in the previous example. The standard MSA with three sources at 10, 3, and 3 representatives each is a more difficult scenario in which to employ the fractional factorial approach. But it is possible to run a study that requires fewer runs. Because the repetition effect can be computed even if there is only one repeat per combination, the easiest way in which to simplify the design is to run a full factorial on just parts and appraisers for a total of $10 \times 3 = 30$ combinations. When there is only a single repeat and the residual ANOVA term is used to estimate error, there is a danger that strong interactions might corrupt the analysis. However, in many practical situations it is found that interaction effects are secondary in impact and can be safely ignored for the sake of reduced sample size. This modified standard MSA design with a single repeat is shown in Table 8.7. Table 8.8 shows the results from the analysis of this modified MSA.

Table 8.7   The full factorial without repeats.

Obs	Part	App	Rep	Meas
1	1	1	1	103.715
2	1	2	1	102.653
3	1	3	1	102.459
4	2	1	1	102.361
5	2	2	1	100.772
6	2	3	1	100.856
7	3	1	1	97.624
8	3	2	1	96.400
9	3	3	1	96.165
10	4	1	1	95.325
11	4	2	1	93.862
12	4	3	1	93.815
13	5	1	1	100.915
14	5	2	1	99.701
15	5	3	1	99.838
16	6	1	1	100.971

(continued)

(continued)

Table 8.7   The full factorial without repeats.

Obs	Part	App	Rep	Meas
17	6	2	1	99.282
18	6	3	1	99.630
19	7	1	1	98.372
20	7	2	1	96.770
21	7	3	1	96.052
22	8	1	1	100.335
23	8	2	1	98.539
24	8	3	1	98.318
25	9	1	1	103.381
26	9	2	1	102.160
27	9	3	1	101.999
28	10	1	1	101.848
29	10	2	1	100.415
30	10	3	1	100.446

Table 8.8   Results of reducing to one repeat in the standard MSA study.

Error Source	Variance Estimate	Standard Deviation	Percent Tolerance
Parts	7.47	2.73	—
Appraiser	0.73	0.85	25.6
Repeatability	0.05	0.22	6.6
RandR	0.78	0.88	26.4

Computer-Generated Designs

It is also possible to use computer-generated designs for MSA studies. Most computer-generated designs start with a full factorial set of combinations and then by a specific search technique arrive at a reduced set of runs that satisfy the investigator's requirements with fewer runs (Myers, 2002). There are some tricks involved with computer-generated designs that should be understood, but this book will not cover these in detail. Just as an improperly chosen fractional factorial can compromise the results, so can some computer-generated designs. It is recommended that if one seeks to use these kinds of designs, one should read some of the references.

Hopefully it is clear that for a large measurement uncertainty study in which one wants to explore and estimate many sources of measurement variation simultaneously, the full factorial is not practical. For example, if there are 10 sources of measurement variation, each with only two levels that need to be considered, the full factorial will require $2^{10} = 1024$ combinations. Almost all of these runs are used to estimate the repeatability effect alone; as a result, the other estimates are poor. However, a fractional factorial could require only 16 combinations. There is still a precision problem, but at least one can get the basic sources without all the runs.

ADEQUATE SAMPLE SIZE ATTAINMENT

Although the methods presented in the first part of this chapter do guarantee separation of the various dispersion sources, they do not by themselves ensure that the estimates will be very reliable. As shown in Chapter 3, the variability of an estimate is really related to two things that the experimenter can choose: the precision of the repeatability estimate and the number of representative elements chosen for each uncertainty source in the study. In the standard MSA study the part component of variability is estimated more consistently because it has 10 parts represented compared to appraiser, which has only three levels. Again, please note that these representative levels are assumed to be randomly chosen from a large set of possible elements. The consistency of the repeatability term is given pretty much by the total number of runs in the study minus the number of levels for the other factors. In the standard MSA study this implies that the repeatability estimate uses roughly $90 - 10 - 3 = 77$ runs from the study and is correspondingly much more consistent.

The Equally Weighted Study

In most circumstances the measurement sources other than repeatability are probably sought to be estimated with more or less equal precision. If this is true, the problem of adequate sample size reduces to two questions. First, how many representative settings are required for any measurement variability source (because they will all be the same), and second, how much of the study should one devote to getting a good estimate of repeatability? Even these two questions are difficult to answer in all generality, but there are some guidelines that can be used. Figure 8.1 shows the relationship between the number of levels and the relative size of the confidence interval for the estimate, assuming a repeatability of 1 and an effect size of 1. In these formulas it is assumed that there are r repeats, p parts, and a appraisers. The $V(\ )$ stands for the estimated variance for each associated component of measurement uncertainty.

$$Variance(V(R)) = \frac{2 * V(R)^2}{ap(r-1)}$$

$$Variance(V(A)) = \frac{2}{(rp)^2} * \left[\frac{(rpV(A)+V(R))^2}{a-1} + \frac{V(R)^2}{ap(r-1)} \right]$$

$$Variance(V(P)) = \frac{2}{(ra)^2} * \left[\frac{(raV(P)+V(R))^2}{p-1} + \frac{V(R)^2}{ap(r-1)} \right]$$

Figure 8.1 Formulas for variances of measurement component estimates.

Because one does not know in advance the values of the variance components that are required for the formulas, one should either take some guesses at their values or use some kind of strategy. One can easily program the values into a spreadsheet and then compute a range of possible design values. The results of such a computer run are shown in Table 8.9. For example, if we wanted the interval to be roughly +/–3 percent around repeatability, +/–50 percent around reproducibility, and +/–100 percent around part, then seven appraisers with 16 parts and two repeats would suffice. Only one repeat would do as well, but the formulas require $r-1 > 0$. Thus $7*16 = 114$ would suffice.

For example, if one has parts and appraisers as in the standard MSA in addition to the repeatability effect, one can look at the table and see that the estimate will be relatively poor if the number of representative levels is less than six. Note that these are relative percentages and that the absolute interval values will depend on the actual estimate of repeatability achieved in the study. Notice that even in this design most of the observations are used to ground the estimate of repeatability. If one assumes that there are no strong interactions, then using only a single reading with no additional repetitions is more efficient.

The Unequally Weighted Study

Although the computations behind the numbers in the table are based on equal numbers of repetitions for each combination, this is not necessary. One might choose to do repeats on a randomly selected 10 percent of the combinations or even to repeat only certain combinations that are thought to best describe typical running conditions. An example of such an unequally weighted MSA that has already been discussed was given in Table 8.8.

Table 8.9 The equally weighted study.

Obs	Pctrep	Pctapp	Pctpart	App	Parts	Reps
1	3.65148	60.5532	103.372	5	15	2
2	2.58199	60.5072	103.349	5	15	3
3	3.53553	60.5446	99.867	5	16	2
4	2.50000	60.5014	99.844	5	16	3
5	3.42997	60.5369	96.695	5	17	2
6	2.42536	60.4964	96.674	5	17	3
7	3.33333	60.5302	93.808	5	18	2
8	2.35702	60.4919	93.787	5	18	3
9	3.24443	60.5241	91.165	5	19	2
10	2.29416	60.4878	91.145	5	19	3
11	3.16228	60.5187	88.734	5	20	2
12	2.23607	60.4842	88.714	5	20	3
13	3.33333	54.1604	103.360	6	15	2
14	2.35702	54.1193	103.341	6	15	3
15	3.22749	54.1527	99.855	6	16	2
16	2.28218	54.1141	99.837	6	16	3
17	3.13112	54.1459	96.685	6	17	2
18	2.21404	54.1096	96.667	6	17	3
19	3.04290	54.1398	93.798	6	18	2
20	2.15166	54.1056	93.780	6	18	3
21	2.96174	54.1344	91.155	6	19	2
22	2.09427	54.1020	1.138	6	19	3
23	2.88675	54.1296	88.724	6	20	2
24	2.04124	54.0987	88.707	6	20	3
25	3.08607	49.4415	103.352	7	15	2
26	2.18218	49.4039	103.336	7	15	3
27	2.98807	49.4344	99.848	7	16	2
28	2.11289	49.3992	99.832	7	16	3
29	2.89886	49.4282	96.677	7	17	2
30	2.04980	49.3951	96.662	7	17	3
31	2.81718	49.4227	93.790	7	18	2
32	1.99205	49.3914	93.775	7	18	3

(continued)

(continued)

Table 8.9 The equally weighted study.

Obs	Pctrep	Pctapp	Pctpart	App	Parts	Reps
33	2.74204	49.4177	91.148	7	19	2
34	1.93892	49.3881	91.133	7	19	3
35	2.67261	49.4133	88.717	7	20	2
36	1.88982	49.3851	88.703	7	20	3

Specially Chosen Source Settings

Another option to reduce the size of an MSA is the specific selection of values for the source settings that approximate certain percentiles of the distributions. This turns the analysis into a type of numerical curve-fitting technique. The settings that are chosen provide a few good points on which to base the curve fit. Often this kind of approach is used in tolerance design experiments. Of course, this approximation approach requires foreknowledge of the distributions of the errors from the sources and could likely be done only after a first MSA study has already been completed. Refer to D'Errico (1988) for an example of using this method for capability analysis. It can be adapted for measurement capability or the MSA as well.

DESIGNING FOR THE UNEXPECTED

In all the cases considered so far, there is clearly an assumption that all important sources of measurement error are captured in one or more of the manipulated factors such as parts or appraisers. But it is wise to anticipate that there will be other, unexpected sources of variation that influence the results of an MSA. For example, a study requiring 90 measurements with interchanged appraisers and parts may take an appreciable time to conduct. During this interval of study it is likely that the ambient temperature will change, the appraisers will suffer fatigue, or the parts will deform. These unanticipated effects may be small compared with the explicit factors in an MSA study, but sometimes they may still be influential. When faced with such unknowns, there are two general approaches that can be followed to manage their effects.

Forcing the Unknown Effects into Repeatability

The first approach seeks to force these unknown effects to enter as repeatability error. This means that the pattern of changes in the unknown effects cannot match the patterns of changes that are forced on the explicitly studied effects. For example, if temperature rises linearly during the course of the study, and if the MSA is conducted with all appraiser 1's readings taken first, followed by all of appraiser 2's, and so on, there can be confusion between temperature effects and appraiser. Another possibility is that the part itself wears out during the study, and there is a gradual change between readings. If appraiser 1 always conducts the first measurement, there could be a bias interjected into the analysis.

If an extraneous effect is suspected, one can take definitive actions to avoid this kind of pattern matching, but if the effect is completely unexpected, a different tactic must be applied. One can ensure that patterns have little chance of matching by randomizing the order of the MSA measurements. In the standard MSA study the repetitions cannot be completely randomized. It is impossible to conduct reading 1 before reading 2, and so on. But the part order and the appraiser order can be randomly chosen, as shown in Table 8.10. Such a randomized pattern reduces the chance that a similar pattern will occur in the hidden factors or in other factors. It would require a very strange natural pattern to match Table 8.10 to any great degree.

Table 8.10   An example of a randomized pattern.

Obs	Part	App	Rep	Meas
1	10	1	1	101.937
2	6	1	1	100.560
3	5	3	1	99.239
4	2	3	1	101.212
5	7	2	1	96.288
6	2	1	1	102.629
7	1	1	1	103.715
8	9	1	1	103.235
9	6	3	1	99.339
10	7	3	1	96.767
11	5	1	1	101.301
12	8	2	1	98.555
13	7	1	1	98.000
14	8	1	1	100.305
15	10	2	1	100.214

(continued)

Table 8.10   An example of a randomized pattern.

Obs	Part	App	Rep	Meas
16	4	2	1	93.945
17	9	2	1	100.968
18	1	3	1	102.342
19	3	1	1	98.103
20	3	3	1	96.518
21	6	2	1	99.400
22	2	2	1	100.895
23	3	2	1	96.536
24	5	2	1	99.359
25	4	1	1	95.137
26	9	3	1	101.888
27	1	2	1	102.150
28	8	3	1	98.781
29	10	3	1	100.745
30	4	3	1	94.017
31	1	2	2	102.137
32	1	1	2	104.230
33	2	2	2	101.257
34	5	2	2	99.306
35	7	2	2	96.219
36	5	1	2	101.340
37	2	3	2	101.099
38	6	3	2	99.985
39	5	3	2	99.729
40	7	1	2	97.720
41	9	1	2	102.793
42	8	3	2	98.717
43	2	1	2	102.743
44	10	3	2	100.815
45	3	1	2	98.077
46	9	2	2	101.634
47	8	2	2	98.479
48	6	1	2	100.399

(continued)

Table 8.10 An example of a randomized pattern.

Obs	Part	App	Rep	Meas
49	10	1	2	102.142
50	1	3	2	102.695
51	3	2	2	96.316
52	8	1	2	100.454
53	3	3	2	96.874
54	6	2	2	98.959
55	9	3	2	101.846
56	4	3	2	93.929
57	4	2	2	93.641
58	4	1	2	95.281
59	7	3	2	96.622
60	10	2	2	100.668
61	3	3	3	96.616
62	10	1	3	102.135
63	2	2	3	101.298
64	4	3	3	94.082
65	4	2	3	94.058
66	5	1	3	101.318
67	6	1	3	100.983
68	7	1	3	97.881
69	2	3	3	101.350
70	5	3	3	99.629
71	1	2	3	102.124
72	9	2	3	101.368
73	3	2	3	95.998
74	7	3	3	96.714
75	10	3	3	100.210
76	6	3	3	99.375
77	1	3	3	102.363
78	8	3	3	98.603
79	5	2	3	99.110
80	7	2	3	96.204
81	2	1	3	102.598

(continued)

(continued)

Table 8.10 An example of a randomized pattern.

Obs	Part	App	Rep	Meas
82	9	3	3	101.635
83	4	1	3	95.215
84	8	2	3	97.985
85	10	2	3	100.785
86	3	1	3	97.263
87	1	1	3	103.939
88	8	1	3	100.414
89	6	2	3	99.412
90	9	1	3	102.989

Adding and Identifying Covariates

The second approach to mitigating the effects of unknown sources of measurement variation is to track as many of them as possible during the running of the study. Then, if necessary, one can try to remove their effects in an enlarged ANOVA. In many real MSA studies one should use both of these methods aimed at reducing the effects of unknown sources of measurement variation. That is, one should take care to record other available effects, and one should try to randomize as much as possible against all other effects. Of course, these safety measures are not guaranteed to work in all cases, and one should always be prepared to validate the results with other independent MSA studies and with physical explanations of the various sources of error.

Chapter 8 Take-Home Pay

1. The design of the MSA ensures the valid estimation of error sources.

2. Reducing the number of repetitions is usually an easy way to reduce size.

3. Fractional factorials and orthogonal arrays can drastically reduce the size of the MSA.

4. The number of levels determines the precision of the error estimates.

5. If possible, MSA design should allow discovery of unexpected effects.

9

The Basic Approach to Attribute Measurement Analysis

A lthough it is usually more efficient and more effective to have measurement systems that deal with quantitative variables values, it is a fact that many practical applications produce only discrete values and often only binary readings. These attribute systems are able to provide only a go/no-go type of result. A common example is the case in which a maintenance mechanic applies a feeler gage to a gap setting in a machine. Notice that in this example the actual processing of the information might be quite complex, but it is the outcome that is simple in that either the gap is set correctly or it is not set correctly. In other situations the fundamental measurement is quantitative, but this more complicated result is transformed into a simpler yes/no result. An example of this is the application of a chemical test for percent O_2 that is made inside an analytics laboratory. The actual test might conclude with an estimate of the actual percent O_2 concentration, but then the technician has been instructed to compare this to a threshold and simply record whether the concentration is within specifications. In this kind of scenario one has a choice of analyzing the performance of this gage as either a variable or an attribute gage, although the truest match is the attribute. Figure 9.1 illustrates this second attribute example.

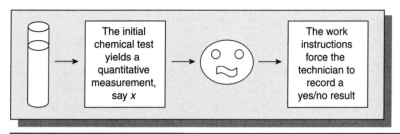

Figure 9.1 A variables measurement becomes an attribute.

THE STANDARD EFFECTIVENESS APPROACH

The Design of the Study

The standard method recommended in the AIAG manual *Measurement Systems Analysis* is a very simple method, which consists of simply counting the number of mistakes made when the attribute measurement system is applied to a sampling of parts. These parts are meant to represent the full range of the process to which the system is applied, and it is recommended that there be at least 50 parts. Each part is presented to each of three operators or inspectors, and a classification is determined at each inspection. Then the same parts are presented a second and third time to the same three operators. It is assumed that the operators are representative of the entire set of possible operators and that the repetitions are made independently of one another. An example of how such a data collection set might look is shown in Table 9.1. It is also assumed that some of these parts, perhaps a third of them, have real product values that the measurement system will find difficult to classify. Some mistakes are not only expected but are required to ensure that the analysis goes properly.

Table 9.1 The standard attribute MSA data collection.

Part	Inspector A			Inspector B			Inspector C			Reference
	Rep 1	Rep 2	Rep 3	Rep 1	Rep 2	Rep 3	Rep 1	Rep 2	Rep 3	
1	Go	Go	Go	Go	Go	Go	Go	Go	Go	Go
2	Go	Go	Go	Go	No-go	Go	Go	Go	Go	Go
3	Go	Go	Go	Go	Go	Go	Go	Go	Go	Go
4	Go	Go	Go	Go	Go	Go	Go	Go	Go	Go
5	No-go	No-go	No-go	No-go	No-go	No-go	No-go	No-go	No-go	No-go
6	Go	Go	Go	Go	Go	Go	Go	Go	Go	Go
7	No-go	No-go	No-go	No-go	No-go	No-go	No-go	No-go	No-go	No-go
8	No-go	No-go	No-go	No-go	No-go	No-go	No-go	No-go	No-go	No-go
9	Go	Go	Go	Go	Go	Go	Go	Go	Go	Go
10	Go	Go	Go	Go	Go	Go	Go	Go	Go	Go
11	Go	Go	Go	Go	Go	Go	Go	Go	Go	Go
12	Go	Go	Go	Go	Go	Go	Go	Go	Go	Go
13	No-go	No-go	No-go	No-go	No-go	No-go	No-go	No-go	No-go	No-go
14	No-go	No-go	No-go	No-go	No-go	No-go	No-go	No-go	No-go	Go
15	Go	Go	Go	Go	Go	Go	No-go	No-go	No-go	Go
16	No-go	No-go	No-go	No-go	No-go	No-go	No-go	No-go	No-go	No-go

(continued)

(continued)

Table 9.1 The standard attribute MSA data collection.

Part	Inspector A			Inspector B			Inspector C			Reference
	Rep 1	Rep 2	Rep 3	Rep 1	Rep 2	Rep 3	Rep 1	Rep 2	Rep 3	
17	No-go	No-go	No-go	No-go	No-go	No-go	No-go	No-go	No-go	No-go
18	Go	Go	No-go	Go	Go	No-go	Go	Go	Go	Go
19	Go	Go	Go	Go	Go	Go	Go	Go	Go	Go
20	Go	Go	Go	Go	Go	Go	Go	Go	Go	Go
21	Go	Go	Go	No-go	No-go	No-go	No-go	Go	No-go	No-go
22	No-go	No-go	No-go	Go	Go	Go	Go	Go	Go	Go
23	Go	Go	Go	Go	Go	Go	Go	Go	Go	Go
24	No-go	No-go	No-go	No-go	No-go	Go	No-go	No-go	No-go	No-go
25	No-go	No-go	No-go	No-go	No-go	No-go	No-go	No-go	No-go	No-go
26	No-go	No-go	No-go	No-go	No-go	No-go	No-go	No-go	No-go	No-go
27	No-go	No-go	No-go	No-go	No-go	No-go	No-go	No-go	No-go	No-go
28	No-go	No-go	No-go	No-go	No-go	No-go	No-go	No-go	No-go	No-go
29	Go	Go	Go	Go	Go	Go	Go	Go	Go	Go
30	Go	Go	Go	No-go	No-go	No-go	No-go	Go	Go	Go
31	Go	Go	Go	Go	Go	Go	Go	Go	Go	Go
32	Go	Go	Go	Go	Go	Go	Go	Go	Go	Go
33	No-go	No-go	No-go	No-go	No-go	No-go	No-go	No-go	No-go	No-go
34	Go	Go	Go	No-go	No-go	No-go	No-go	No-go	No-go	No-go
35	No-go	No-go	No-go	No-go	No-go	No-go	Go	Go	Go	No-go
36	Go	Go	Go	Go	Go	Go	Go	Go	Go	Go
37	No-go	No-go	No-go	Go	No-go	No-go	No-go	No-go	No-go	No-go
38	No-go	Go	Go	Go	Go	Go	Go	Go	Go	Go
39	Go	Go	Go	No-go	No-go	No-go	Go	Go	Go	Go
40	No-go	No-go	No-go	No-go	No-go	No-go	No-go	No-go	No-go	No-go
41	No-go	No-go	No-go	No-go	No-go	No-go	No-go	No-go	No-go	No-go
42	Go	Go	Go	No-go	No-go	Go	No-go	No-go	No-go	No-go
43	Go	Go	Go	Go	Go	Go	Go	Go	Go	Go
44	No-go	No-go	No-go	No-go	No-go	No-go	No-go	Go	No-go	No-go
45	Go	Go	Go	Go	Go	Go	Go	Go	Go	Go
46	No-go	No-go	No-go	No-go	No-go	No-go	No-go	No-go	No-go	No-go
47	No-go	No-go	No-go	No-go	No-go	No-go	No-go	No-go	No-go	No-go
48	No-go	No-go	No-go	Go	No-go	No-go	No-go	No-go	No-go	No-go
49	Go	Go	Go	Go	Go	Go	No-go	Go	Go	Go
50	No-go	No-go	No-go	No-go	No-go	No-go	No-go	No-go	No-go	No-go

The Standard Analysis

The last column in the table represents a standard or reference classification that can be used as truth if it is available. Often this is the judgment of an expert inspector who looks at the same parts. The AIAG manual covers analysis methods for situations in which there is no reference as well as for situations with a reference classification. Only the analysis that includes the reference will be discussed here. This analysis consists of counting the number of mistakes made in these nine repeat readings of the 50 parts. Mistakes occur whenever the inspector reading does not match the reference classification. Specifically, there are four types of important outcomes that can occur for each unique part presented for measurement:

1. Agreement of inspector with reference over all three repeats

2. Three go readings by the inspector with a no-go standard— leak rate

3. Three no-go readings by the inspector with a go standard— false alarm

4. Mixture of go and no-go by the inspector

The effectiveness of an operator's use of the attribute gage is measured by counting the number of parts in which there is no disagreement with the reference on any reading, divided by the total number of parts used in the study. Thus the effectiveness will decrease with every false alarm, leak, and other mixed result that should occur. In addition, the number of false alarms per part and the number of leaks per part are evaluated. Table 9.2 shows the results of these computations for the data presented earlier. Table 9.3 details the requirements for acceptability that are often used.

This approach finds the attribute gage to be unacceptable if there are more than five aberrant parts, more than one leaked part, or more than two false alarms in the standard 50 parts per operator. Note that fewer parts can make these requirements even tighter in the sense that fewer disagreements will lead to an unacceptable result. For instance, a study with only 20 parts would mean that two aberrant parts, one false alarm, or one leak would lead

Table 9.2 The summary analysis for the standard attribute MSA.

	Inspector A	Inspector B	Inspector C
Effectiveness	43/50 or 86%	42/50 or 84%	43/50 or 86%
False Alarm Rate	2/50 or 4%	3/50 or 6%	2/50 or 4%
Leak Rate	3/50 or 6%	0/50 or 0%	1/50 or 2%

Table 9.3 The standard attribute MSA targets for acceptability.

	Acceptable	Marginally Acceptable	Unacceptable
Effectiveness	90%	80%	<80%
False Alarm Rate	5%	10%	>10%
Leak Rate	2%	5%	>5%

to the conclusion that the gage is inadequate. Just as the issue of adequate sample size came into play with regard to the variable MSA in the early chapters of this book, so, too, is it critical for attribute analyses.

SAMPLE SIZE EFFECTS FOR THE ATTRIBUTE ANALYSIS

Real Measurement Error Rates

The results of the attribute MSA are impacted by the sample size of the study. To understand this impact, consider the situation in which the true effectiveness of the true measurement error is 90 percent. That is, on average, if the operator is presented with either a go or a no-go part, there is a 90 percent chance that he or she will agree with the reference value, and there is a 10 percent chance that he or she will disagree. It is assumed that each and every individual reading will be subjected to these errors independently. If there are 150 total opportunities or readings per operator, the chance of zero errors in any of these 50 readings is $.90^{150} = 0.000000137$, which is effectively zero. These raw errors are not used directly in the MSA, because one only counts parts. For example, a part with two misreads counts only one part under this plan. So the raw 90 percent chance of a correct reading is transformed to the chance that three repeats will have zero misreads, which is 0.729. Assuming an underlying true misread rate of 10 percent means that there is a 72.9 percent chance that any part will have a perfect agreement with the reference value. So what is really of interest is the chance that there will be five or fewer misreads of parts where there are 50 parts, and the chance of a misread of a part is .271. This chance of five or fewer disagreements on 50 parts (150 trials) with 90 percent effectiveness is 0.0028. For a study under the same effectiveness that is performed on only 20 parts, the probability of being acceptable is 0.0623. So it is more likely to pass the test when fewer parts are utilized, but it is unlikely in either case to pass the effectiveness test as given with a true 90 percent misread rate. Consider Table 9.4, which lists some of the probabilities for showing effectiveness when the true effectiveness is 90 percent under various sample sizes.

In examining this table, it is clear that as few as 20 parts can lead to unwelcome surprises. At 10 parts there is a good chance (~80 percent) that a system with a base misread rate of 10 percent will be deemed unacceptable. This same probability is only 0.3 percent under a typical scenario of 50 parts.

Targets for Measurement Error Rates

Another way in which one can approach this sample size effect is to reverse the question and ask what the real effectiveness of the measurement system has to be in order to ensure a particular probability of passing the test. If one arbitrarily chooses a 90 percent chance of passing, one might get the results shown in Table 9.5, which gives the minimum true effectiveness for various sample sizes.

Table 9.4 Effect of sample size on effectiveness acceptability.

True	Size	Probability	True	Size	Probability	True	Size	Probability
99%	10	96.612%	90%	10	19.998%	80%	10	1.304%
99%	20	97.954%	90%	20	6.234%	80%	20	0.030%
99%	50	99.644%	90%	50	0.282%	80%	50	0.000%
99%	100	99.980%	90%	100	0.002%	80%	100	0.000%
99%	150	99.999%	90%	150	0.000%	80%	150	0.000%

Table 9.5 Minimum true effectiveness required to ensure 90 percent probability of passing.

Probability of Passing	Sample Size	Required True Effectiveness
90%	10	98.2%
90%	20	98.1%
90%	50	97.9%
90%	100	97.6%
90%	150	97.5%

THE EFFECTS OF NONINDEPENDENCE IN THE REPEATS

Along with sheer sample size effects, the distribution of samples in an attribute study can have a significant impact on the results. For example, it is very difficult to guarantee that every evaluation an inspector makes on a part is independent of the other repeats on that part. This is difficult to maintain even for variables measurements, but for attribute measurements it is often nearly impossible. This is because the features that attribute measurement systems treat are often visual and can easily be remembered by good inspectors. When the repeat inspections are not independent, even if they are not completely correlated, the whole probability structure can change and the guarantee afforded by the MSA study can deteriorate rapidly.

As an example, consider the simplistic case in which the inspector remembers his original classification on a part and simply duplicates this result for the other two repeats on each part. It is clear that this effectively reduces the sample size to 50 per operator rather than 150. Or in terms of the probability of all three repeats agreeing with the reference, this becomes 90 percent, rather than 72.9 percent as earlier. A quick computation shows that the chance of passing the acceptability criterion is now the chance of getting five or fewer misreads with a base probability of .10, which is 62.612 percent.

In addition to effectiveness calculations, false alarms are judged by the number of parts in which all three repeats differ from the reference classification of go. In this new, nonindependent scenario this will happen every time that the first reading is no-go when the reference is go. The chance of three independent classifications reading the same but different from the reference in this fashion is much less than the chance of a single difference. If the actual false alarm rate is 1 percent, the chance of three false alarms on one part is .01*.01*.01 = .000001. So the guarantee given in the case in which the observations are not independent is quite different from the standard attribute MSA design. Of course, a similar problem occurs with the leak rate estimation. Because the requirements for the leak rate are 2 percent for the attribute gage to be acceptable, the problem becomes even more serious. That is, the 2 percent requirements are really looser than that nominal value for the standard attribute MSA study, and the nonindependent case is much more severe. And similarly, this converts to differences in basic false alarm rate and in the effects of sample size.

EFFECTS OF POORLY CHOSEN REFERENCE PARTS

The standard attribute MSA suggests that the parts be chosen to cover the range of values characteristic of the process to which the measurement is applied. Furthermore, it is often suggested that roughly one-third of the parts have true values far outside the tolerances, one-third have parts near the target, and the remaining one-third have values that border the tolerance limits themselves. Observe Figure 9.2, which seeks to illustrate this approach to part selection.

A problem can occur with the standard attribute measurement system study if these parts are chosen differently from the prescribed method given earlier. A poor selection can be accomplished in at least three ways:

1. Only clearly go and no-go parts are chosen with no indeterminate ones.

2. Only indeterminate parts are chosen with no clear cases.

3. Differing fractions of the various parts are chosen.

A scenario following possibility number one in the list could have the 50 parts coming from the extremes of the specification range. These true part values are so far from the tolerance that there may be little or no difficulty in classifying them. It is easy to see in this case that all estimates of error will be zero. That is, the gage will have 100 percent effectiveness, zero false alarms, and zero leaks. It is certainly possible for the measurement system to be so excellent that it has no errors, but in this case the study is

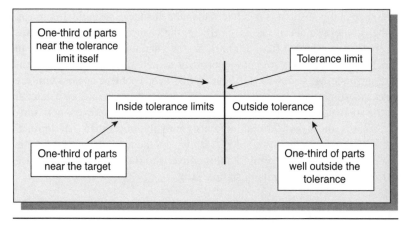

Figure 9.2   The positioning of part values for the attribute MSA.

faulty and simply does not allow a fair determination of the error. This kind of mistake in the part selection may make the system look far better than it really is, and that could lead to costly errors in application. This kind of mistake seems simple enough to avoid, but in some practical situations it takes a good deal of effort to do so.

A second kind of part selection mistake can occur when one considers only parts right near the specification boundaries. In this case the slightest measurement error can create false alarms and leaks. At first sight this seems to be an ideal condition if one wants to estimate these error rates in the most efficient way. But this is not the way in which the AIAG standard describes the MSA study. It recommends that the entire range of parts be represented in the selection. Inherent in this choice is the assumption that error rates can depend on the actual value of the part. That is, it is assumed that the leak rate is near zero well outside the tolerance range and then increases as the part value nears the tolerance. Finally, when the true part value crosses the tolerance, by definition there can be no false alarm rate at least until the true part value crosses the other tolerance limit. And, vice versa, there cannot be a leak rate if the true part value is located within the specifications. In a similar but reflected way the leak rate probability rises from near zero at the center of tolerances—that is, usually at the process target to a maximum at the tolerances themselves. Consider Figure 9.3 to gain a graphical understanding of this relationship between the true part value and the two classification error types.

This logic of part selection assumes that the resulting effectiveness computation can be a mixture of measurement error probabilities that are achieved when the system is applied to this distribution of part values. As an

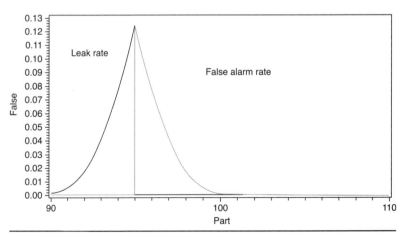

Figure 9.3 The expected error rates compared to part tolerances.

example, assume the conditions shown in Table 9.6. The basic false alarm rates and leak rates are given for the parts chosen outside the tolerances, within the tolerances, and near the tolerances.

When the standard attribute MSA is conducted with parts coming from these four regions, clearly the overall estimates of effectiveness, leak rate, and false alarm rate will be affected. One might begin with a look at leak rate, which is estimated by the number of parts in which all three repeats have measurements that are classified as go when the reference classification is no-go. In the split-thirds approach it is clear that a third of the parts chosen well outside the tolerances have a low (in this example, zero) chance of producing a leak. Another third of the parts have true values located well inside the tolerances, and by definition these can have no leak rate at all. Of the third of the parts near the tolerance itself, roughly half will be just inside the tolerance, so again, they cannot have a leak rate. So, one-half of one-third of the 50 parts exercise the leak rate mechanism at a serious level. Of the 50 parts specified in the standard attribute MSA, only about eight really exercise the measurement error leak rate.

Turning the examination to false alarm rate, it can be seen that the same sort of thing affects its estimate as well. That is, only about eight parts really exercise the false alarm rate as well. As for effectiveness, because it depends on the total of parts that experience false alarm error or leak error, it exercises only about 16 parts. With these restrictions in mind, one can see that the entire study really reduces to the information in the 16 parts. The other 34 parts play no role in most attribute measurement systems if this sample selection plan is used. Consider the example data given in Table 9.7. This table is based on the same data that were presented in Table 9.1 to represent a standard attribute MSA study but simply relates for each part in which zone its true value lies. The table also has a column indicating the number of errors each operator made. One can clearly see from this table that the only errors occur in the parts near the tolerance limits.

Table 9.6 An example of three different error rates.

True Part Value Location	False Alarm Rate	Leak Rate
Well inside tolerances	0	0
Well outside tolerances	0	0
Just inside tolerances	0	0.05
Just outside tolerances	0.10	0

Table 9.7 The part zones of Table 9.1.

Part	Zone	A	B	C	Ref	Part	Zone	A	B	C	Ref
1	Well in	0	0	0	Go	26	Well out	0	0	0	No-go
2	Just in	0	1	0	Go	27	Just out	0	0	0	No-go
3	Well in	0	0	0	Go	28	Well out	0	0	0	No-go
4	Well in	0	0	0	Go	29	Well in	0	0	0	Go
5	Well out	0	0	0	No-go	30	Just in	0	3	1	Go
6	Well in	0	0	0	Go	31	Well in	0	0	0	Go
7	Well out	0	0	0	No-go	32	Well in	0	0	0	Go
8	Well out	0	0	0	No-go	33	Well out	0	0	0	No-go
9	Just in	0	0	0	Go	34	Just out	3	0	0	No-go
10	Well in	0	0	0	Go	35	Well out	0	0	3	No-go
11	Well in	0	0	0	Go	36	Well in	0	0	0	Go
12	Just in	0	0	0	Go	37	Just out	0	1	0	No-go
13	Just out	0	0	0	No-go	38	Just in	1	0	0	Go
14	Just in	3	3	3	Go	39	Just in	0	3	0	Go
15	Well in	0	0	3	Go	40	Well out	0	0	0	No-go
16	Well out	0	0	0	No-go	41	Well out	0	0	0	No-go
17	Well out	0	0	0	No-go	42	Just out	3	1	0	No-go
18	Just in	1	1	0	Go	43	Well in	0	0	0	Go
19	Well in	0	0	0	Go	44	Just out	0	0	1	No-go
20	Well in	0	0	0	Go	45	Well in	0	0	0	Go
21	Just out	3	0	0	No-go	46	Well out	0	0	0	No-go
22	Just in	3	0	0	Go	47	Well out	0	0	0	No-go
23	Well in	0	0	0	Go	48	Just out	0	1	0	No-go
24	Just out	0	1	1	No-go	49	Just in	0	0	1	Go
25	Well out	0	0	0	No-go	50	Well out	0	0	0	No-go

The false alarm rate, the leak rate, and the effectiveness results for each operator have already been presented in Table 9.2. But the same results are achieved in this example only if the 16 parts that are near the tolerance are examined. This example is made especially to demonstrate the effects of sample value distribution in the attribute MSA study, and so the impact is clear. In more realistic studies it is harder to know where the part true values lie, so this impact, although still present, is harder to evaluate. If one extracts only the 16 essential parts from the full table and computes the same rates relevant, one must also alter the effectiveness targets. It must be emphasized that there is nothing inherently wrong with the AIAG approach as a baseline, although one might argue that the study could be run more efficiently than proposed. The critical criterion is to correctly align the study with the acceptance criteria.

CORRECTIONS TO THE PROCEDURE

The AIAG manual does offer another choice that corrects some of the problems that have been exposed by the discussion. It is called the *analytic method,* and although it seems not to be overly popular, it should be considered as a viable alternative. This method fundamentally consists of selecting parts along the whole range of true parts values, ideally at 10 equidistant points along this range. Figure 9.4 shows what this distribution might look like. It is then recommended that 10 repeats be made at these 10 points along the scale. It does not discuss having several operators involved in the study. Thus one can assume that there are 100 points at 10 parts in this approach.

The numbers of repeats classified as go are then computed for each part—that is, each of the points in the range. Assuming the same kind of error curve dependent on the true part value, one should see the proportion starting low or at zero and then rising until the tolerance is reached, at which point it should start to fall again until it reaches zero at the center of the tolerance range.

In this approach one essentially converts or transforms the attribute character of the measurement to that of the variable measurement. This

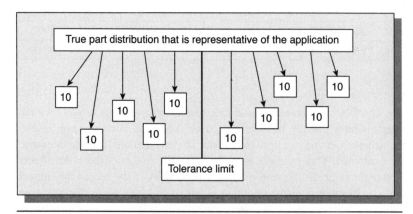

Figure 9.4 Distribution of part values for the analytic procedure.

transformation is only approximate because each proportion is based on only 10 results, but it does serve the purpose of essentially turning the study into a variable gage study. This is both an advantage and a disadvantage. It is an advantage to use the variable approach because one can more readily estimate the bias and repeatability of the process measurement, and it is less sensitive to the sample size effects. But it is also a disadvantage because now the requirements that pertain are those of a variables gage, which have been shown to be more stringent than those for an attribute gage. The advantages probably outweigh the disadvantages in most cases. This is especially true if one wants to extend the analysis to other factors, more levels, and other more powerful analyses, as shown in Chapter 10.

The attribute method is also very sensitive to the number and values of easily identified parts that are chosen for the study. Because the scale for the fitting of the normal distribution is stretched by these values, it is quite possible to change the measurement system results dramatically by simply changing the value of one of the two true part values. A similar effect can be introduced by unequal spacing in the interior true values. Again, it should be emphasized that all these difficulties can be overcome with care, but there is no correction available when the measurement study is not performed at all.

Chapter 9 Take-Home Pay

1. Standard attribute MSA studies estimate errors by simple counts.

2. The selection of parts is critical and also wasteful in many attribute studies.

3. The method of counting errors can be misleading.

4. The standard method does not work effectively with error rates that vary with part value.

10

A More Powerful Approach to Attribute Measurement Analysis

B ecause of the various potential problems that were detailed in Chapter 9, it is desirable to use a better, more reliable, more flexible method for the analysis of attribute measurement systems. The method recommended here is an extension of the analytic method and is described completely in Boyles (2001). It employs a fairly sophisticated use of maximum likelihood estimation for the measurement error model. It applies this analysis in an approach that iterates between two steps of estimation and expectation until a final answer is produced. This method also allows the computation of confidence intervals, but this feature is not utilized in the applications of the methods described in this book. There are four variants of this method that will be presented:

1. Repeatability for a single operator without a reference

2. Repeatability for a single operator with a reference

3. Repeatability for multiple operators with a reference

4. Modification and extensions to more effects

REPEATABILITY WITH NO REFERENCE

In this first application scenario there is a single operator without any reference values. Consider a study in which parts are chosen to represent the range of values of interest in the application. Each part, possibly 20 of them, will have a number of repeat inspections, perhaps 10. Of these 200 results there will be some go values and some no-go values. Each part will thus have a percentage or fraction of go and no-go values. Fundamentally, these fractions allow the analysis to proceed. Assuming that the error rate for misclassifying a bad part is smaller than the error rate for misclassifying a good part, it is possible to estimate three probabilities from these kinds of data. These three probabilities are as follows:

1. $p(go)$ = the probability of go values in the set of test values

2. $p(no\text{-}go|go)$ = the probability of classifying a good part as bad

3. $p(go|go)$ = the probability of classifying a good part as good

Notice that these probabilities allow one to derive the probabilities of classifying a bad part as bad and classifying a bad part as good. Consider Table 10.1, which shows an example set of balanced data in which there are 20 parts and 10 repeats of each part, for a total of 200 measurements made. In this example it is assumed that there is no known or reference value for each part. Notice that some parts are consistently characterized as nonconform in every repeat, so they likely correspond to parts whose true value is

Table 10.1 The data for a single operator attribute study without a reference.

Part	Rep 1	Rep 2	Rep 3	Rep 4	Rep 5	Rep 6	Rep 7	Rep 8	Rep 9	Rep 10	Go Total
1	Go	Go	Go	No-go	No-go	No-go	No-go	No-go	No-go	No-go	3
2	Go	No-go	No-go	Go	Go	Go	Go	Go	Go	Go	8
3	Go	Go	Go	Go	Go	Go	Go	Go	Go	Go	10
4	Go	No-go	Go	No-go	Go	Go	No-go	Go	Go	Go	7
5	No-go	No-go	No-go	No-go	No-go	No-go	No-go	No-go	No-go	No-go	0
6	No-go	No-go	No-go	No-go	No-go	No-go	No-go	No-go	No-go	No-go	0
7	No-go	No-go	No-go	No-go	No-go	No-go	No-go	No-go	No-go	No-go	0
8	Go	Go	Go	Go	Go	Go	Go	Go	Go	Go	10
9	Go	No-go	Go	Go	No-go	Go	Go	No-go	No-go	Go	6
10	Go	Go	No-go	No-go	No-go	No-go	No-go	No-go	No-go	No-go	2
11	No-go	No-go	No-go	No-go	No-go	No-go	No-go	No-go	No-go	No-go	0
12	No-go	No-go	No-go	No-go	No-go	No-go	No-go	No-go	No-go	No-go	0
13	No-go	No-go	No-go	No-go	No-go	No-go	No-go	No-go	No-go	No-go	0
14	Go	Go	Go	Go	Go	Go	Go	Go	Go	Go	10
15	Go	Go	Go	Go	Go	Go	Go	Go	Go	Go	10
16	No-go	No-go	Go	Go	No-go	No-go	No-go	No-go	No-go	No-go	2
17	No-go	No-go	No-go	No-go	No-go	No-go	No-go	No-go	No-go	No-go	0
18	No-go	No-go	No-go	No-go	No-go	No-go	No-go	No-go	No-go	No-go	0
19	Go	Go	Go	Go	Go	Go	Go	Go	No-go	Go	9
20	No-go	No-go	No-go	No-go	Go	Go	No-go	Go	No-go	No-go	4

well outside the specification limits. And other parts are consistently classified as conform in every repeat, so their true values are likely well within the
specifications. The other parts have a mixed set of results in which there are
conforming and nonconforming classifications on the same parts. These are
parts whose true values likely fall near the specification limits.

This method produces the estimates given in Table 10.2 for the example
data presented in Table 10.1. The probability p(no-go) is not shown in the
table, but it could be computed as $1 - p$(go).

These estimates tell one directly the ability of the device to make a correct determination when presented with either a true go or a true no-go part.
The estimate of the p(go) depends on the specific set of parts chosen for the
study. If this choice is not representative of the real application, it might be
valuable to evaluate the risk by combining these measurement results with
an estimate of p(go) from the representative process for an arbitrary 1000
parts. Or one could choose a whole range of hypothetical values to see the
range of risks as a sequence of possible percentages, as shown in Table 10.3.
This could easily be displayed as a curve, as shown in Figure 10.1.

Table 10.2 Estimates of probabilities for the one-operator study without a reference.

Probability	Estimated Value
p(go)	0.40186
p(go|go)	0.87267
p(go|no-go)	0.090798
p(no-go|go) = 1 − p(go|go)	1 − 0.87267 = 0.12733

Table 10.3 A set of hypothetical p(go) values and the risks they generate.

Hypothetical p(go)	Hypothetical p(no-go)	p(no-go|go)	Parts Falsely Scrapped	p(go|no-go)	Parts Falsely Leaked
.50	.50	0.13	65	0.09	45
.60	.40	0.13	78	0.09	54
.70	.30	0.13	91	0.09	63
.80	.20	0.13	104	0.09	72
.90	.10	0.13	117	0.09	81
.95	.05	0.13	123	0.09	85
.99	.01	0.13	129	0.09	89
.999	.001	0.13	130	0.09	90

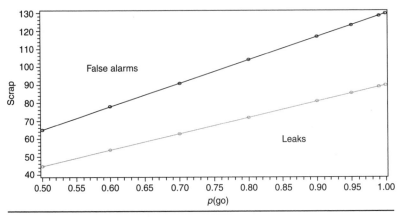

Figure 10.1   A graph of the risks due to misclassification errors.

ADDING COSTS TO THE ANALYSIS

Although one is often forced to make decisions on the basis of error rates, it is usually better to use actual costs of misclassification if they are available. For example, there might be a cost that is incurred each time a false alarm occurs—that is, whenever a go part is measured as a no-go. In a worst-case scenario, the entire part might be scrapped if the measurement shows that it is out of specification. The cost in this case could be set identical to the manufacturing cost or set equal to the sale price if no new good unit can be substituted in the shipment in time. More often than not, companies are used to making nonconforming parts and might have established procedures to recoup the material, sell it at a reduced price, or repair the part. Each of these actions has a cost associated with it that is, hopefully, less than the full manufacturing cost. Whatever this cost is, it can be used to establish an expected cost due to the measurement system when it is applied to a real process.

The second cost is incurred when a no-go part is mismeasured as a go part and ultimately shipped to the customer. This cost is usually larger than the cost of a false alarm. It can be very large if the leaked part leads to poor performance in the customer's hands. Even if there are additional inspection systems that prevent the worst leaks, there is an extra cost incurred for these quality checks. In the example, assume that the cost of a false alarm is $5 per part, and the cost of a leak is $25 per part. These costs can be combined with the data in Table 10.3 to compute expected dollar losses for the hypothetical 1000 parts. These cost data are added to Table 10.3 and given as Table 10.4.

Table 10.4   Potential costs due to measurement errors in a real application.

Hypothetical p(go)	Hypothetical p(no-go)	p(no-go|go)	Parts Falsely Scrapped	Cost of False Alarms	p(go|no-go)	Parts Falsely Leaked	Cost of Leaks	Total Costs
.50	.50	0.13	65	$325	0.09	45	$1125	$1450
.60	.40	0.13	78	$390	0.09	54	$1350	$1740
.70	.30	0.13	91	$455	0.09	63	$1575	$2030
.80	.20	0.13	104	$520	0.09	72	$1800	$2320
.90	.10	0.13	117	$585	0.09	81	$2025	$2610
.95	.05	0.13	123	$615	0.09	85	$2125	$2740
.99	.01	0.13	129	$645	0.09	89	$2225	$2870
.999	.001	0.13	130	$650	0.09	90	$2250	$2900

USING PREDICTION PROBABILITIES

The method recommended here generates a predicted value for each and every part as well as the estimated probabilities of interest. The validity of these predicted values depends on the number of parts, the observed number of go decisions, and the number of repetitions used. In the example, if one uses 20 iterations, one might get the predicted probabilities of being go values that are given in Table 10.5 for the example data. The algorithm classifies the final disposition of the part as go or no-go depending on this predicted value. If this prediction is less than 0.50, the part is classified as a go part; otherwise it is classified as a no-go part in the standard procedure.

Table 10.5   The predicted values of the data points and their ultimate classification.

Part Number	Predicted Probability of Go	Final Classification Based on .50
1	0.00063	No-go
2	1.00000	Go
3	1.00000	Go
4	0.99993	Go
5	0.00000	No-go
6	0.00000	No-go

(continued)

(continued)

Table 10.5 The predicted values of the data points and their ultimate classification.

Part Number	Predicted Probability of Go	Final Classification Based on .50
7	0.00000	No-go
8	1.00000	Go
9	0.99511	Go
10	0.00000	No-go
11	0.00000	No-go
12	0.00000	No-go
13	0.00000	No-go
14	1.00000	Go
15	1.00000	Go
16	0.00001	No-go
17	0.00000	No-go
18	0.00000	No-go
19	1.00000	Go
20	0.04145	No-go

DIFFERENT DECISION RULES

The decision rule of 0.50 is correct for most circumstances, but one might want to alter it to take into account the different costs between false alarms and leaks. If the costs are $25/$5 = 5 to 1 ratio, as in the example, one might want to bias the final classification toward reducing leaks at the expense of more false alarms. The data in Table 10.5 would require a substantial change from the 0.50 rule to make any difference in the final classification in this example. But if the predictions were different, as given in Table 10.6, then one might want to use a different cutoff to decide the final classifications. A cutoff of 0.20 would force more classifications as no-go values and therefore reduce the number of leaked parts. Table 10.6 also shows the classifications using 0.50 and 0.20.

Table 10.6 Applying a different set of cutoffs to classify parts.

Part Number	Predicted p(go)	Classification Based on 0.50	Classification Based on 0.80
1	1.00000	Go	Go
2	0.60611	Go	No-go
3	0.12080	No-go	No-go
4	0.00000	No-go	No-go
5	0.86005	Go	Go
6	0.3775	No-go	No-go
7	0.00000	No-go	No-go
8	1.00000	Go	Go
9	0.15601	No-go	No-go
10	0.65601	Go	No-go

REPEATABILITY WITH A REFERENCE

An immediate extension of the previous approach to attribute measurement systems is to add a reference value for each part. This reference value can be obtained by expert opinion, by measurement with a less error-prone system, or by using an average of multiple readings. Generally, this reference value is better, but it too can have some measurement error associated with it. Often in the practical situations in which attribute measurement systems are applied, the reference is really a reference method or a person who is simply better trained or more experienced than the normal appraiser. An appreciable error in the reference determination itself can have quite dramatic effects on the final measurement performance assessment. The method presented here can take the reference error into account.

Consider the example data given in Table 10.7, in which a reference value is given, assumed to be error free. That is, the $p(go|go) = 1$ and $p(no-go|no-go) = 1$ for the reference system in this initial example. The fitted probabilities are given in Table 10.8.

Now observe the data and results if the reference measurement system also has some error associated with it. Although the expectation is that a true

Table 10.7 The one-operator system with perfect reference values.

Part Number	Reference Class	Number of Repeats	Number of Go Values
1	No-go	10	3
2	Go	10	8
3	Go	10	10
4	Go	10	7
5	No-go	10	0
6	No-go	10	0
7	No-go	10	0
8	Go	10	10
9	No-go	10	6
10	No-go	10	2
11	No-go	10	0
12	No-go	10	0
13	No-go	10	0
14	Go	10	10
15	Go	10	10
16	No-go	10	2
17	No-go	10	0
18	No-go	10	0
19	Go	10	9
20	No-go	10	4

Table 10.8 The fitted probabilities for the perfect reference example.

Probability	Estimated Value
$p(go)$	0.40000
$p(go|go)$	0.75000
$p(go|no\text{-}go)$	0.08333
$p(no\text{-}go|go) = 1 - p(go|go)$	$1 - 0.75000 = 0.25000$

reference system might have almost perfect performance, in this example the $p(go|go)$ will be taken as 0.80, and the $p(go|no\text{-}go) = 0.05$ for this reference system. The new data are identical to that shown in Table 10.7, but the new results are shown in Table 10.9.

Table 10.9   The fitted probabilities for the imperfect reference example.

Probability	Estimated Value
$p(\text{go})$	0.43038
$p(\text{go}|\text{go})$	0.69706
$p(\text{go}|\text{no-go})$	0.08778
$p(\text{no-go}|\text{go}) = 1 - p(\text{go}|\text{go})$	$1 - 0.69778 = 0.30222$

ADDING REPRODUCIBILITY EFFECTS

It is also possible to extend the method to include varying error rates for each of multiple operators. The comparison of the results over the three operators can lead to insight and correction of problems. Refer to Boyles (2001) for the mathematical details. Consider an example in which there are 20 parts with 10 repeat measurements for each part determined by three different operators with a reference value that has some error associated with it. When the reference value is used, it is assumed the error probabilities associated with this standard are $p(\text{nc}|\text{c}) = .05$ and $p(\text{c}|\text{nc}) = .02$. Table 10.10 shows the data, and Table 10.11 shows the estimated probabilities under the different scenarios for each operator. In this example there are a few parts that are classified differently by one operator when compared to the other two operators. This is a difference that one can explore with the aim of eliminating or reducing this discrepancy. It is also possible to modify the method to allow for a single estimate of the $p(\text{go})$ term, ensuring that it does not change from operator to operator.

Table 10.10   The data for three operators with a reference.

Part Number	Operator A		Operator B		Operator C		Reference Class
	Number of Measures	Number of Go Values	Number of Measures	Number of Go Values	Number of Measures	Number of Go Values	
1	10	3	10	0	10	6	No-go
2	10	8	10	7	10	6	Go
3	10	10	10	7	10	10	Go
4	10	7	10	7	10	10	Go
5	10	0	10	0	10	6	No-go
6	10	0	10	0	10	6	No-go

(continued)

(continued)

Table 10.10 The data for three operators with a reference.

Part Number	Operator A Number of Measures	Operator A Number of Go Values	Operator B Number of Measures	Operator B Number of Go Values	Operator C Number of Measures	Operator C Number of Go Values	Reference Class
7	10	0	10	2	10	2	No-go
8	10	10	10	8	10	9	Go
9	10	6	10	4	10	0	No-go
10	10	2	10	4	10	0	No-go
11	10	0	10	2	10	0	No-go
12	10	0	10	2	10	7	No-go
13	10	0	10	0	10	0	No-go
14	10	10	10	10	10	9	Go
15	10	10	10	8	10	9	Go
16	10	2	10	4	10	4	No-go
17	10	0	10	0	10	0	No-go
18	10	0	10	0	10	2	No-go
19	10	9	10	9	10	4	Go
20	10	4	10	4	10	4	No-go

Table 10.11 Results for three operators with a reference.

Probability	Operator A Estimate	Operator B Estimate	Operator C Estimate
$p(go)$	0.40081	0.55654	0.48605
$p(go|go)$	0.74849	0.44921	0.57608
$p(go|no\text{-}go)$	0.09345	0.11275	0.07783
$p(no\text{-}go|go)$	0.25151	0.55079	0.42392

OTHER EXTENSIONS AND ISSUES

This method can also be extended to cover other interesting situations. For example, following the lead of Chapter 4 it might be interesting to include other factors in the study, such as setup, location, or time period. The model can be extended to include these effects in a similar way to that used for operator effects. For example, time periods may be quickly substituted for operators in the analysis. Or, both operators and time periods may be introduced. The easiest way to accommodate these factors is to treat them in combination and compute the probabilities for each unique combination. For example, consider the data given in Table 10.12, which has five parts, 10 repeats for the nine combinations of three operators over three time periods. This is analyzed as if there were a single factor, operator × time period, with nine values. The results are shown in Table 10.13. In some situations it might be interesting to then average the probabilities over the results to get a result by operator or by time period. This secondary analysis is shown in Table 10.14.

Table 10.12 An example with operators and time periods.

Part	Period	Operator A Measures	Operator A Total Go	Operator B Measures	Operator B Total Go	Operator C Measures	Operator C Total Go	Reference
1	I	10	0	10	0	10	6	No-go
2	I	10	2	10	4	10	4	No-go
3	I	10	8	10	5	10	8	Go
4	I	10	9	10	5	10	5	Go
5	I	10	10	10	10	10	10	Go
6	I	10	10	10	7	10	7	Go
7	I	10	5	10	2	10	0	No-go
8	I	10	5	10	2	10	0	No-go
9	I	10	2	10	1	10	1	No-go
10	I	10	1	10	0	10	6	No-go
1	II	8	0	10	4	6	2	No-go
2	II	8	3	10	4	6	4	No-go
3	II	8	8	10	2	6	6	Go
4	II	8	8	10	8	6	5	Go
5	II	8	8	10	10	6	6	Go
6	II	8	6	10	7	6	6	Go
7	II	8	3	10	2	6	0	No-go
8	II	8	0	10	2	6	0	No-go
9	II	8	2	10	5	6	3	No-go
10	II	8	1	10	2	6	5	No-go
1	III	6	0	8	0	10	2	No-go
2	III	6	2	8	4	10	4	No-go

(continued)

(continued)

Table 10.12   An example with operators and time periods.

Part	Period	Operator A		Operator B		Operator C		Reference
		Measures	**Total Go**	**Measures**	**Total Go**	**Measures**	**Total Go**	
3	III	6	5	8	5	10	8	Go
4	III	6	6	8	5	10	8	Go
5	III	6	5	8	8	10	10	Go
6	III	6	6	8	7	10	7	Go
7	III	6	5	8	1	10	3	No-go
8	III	6	3	8	2	10	0	No-go
9	III	6	1	8	3	10	1	No-go
10	III	6	1	8	1	10	3	No-go

Table 10.13   The results of the study with operators and time periods.

Probability	Time Period	Operator A	Operator B	Operator C
$p(go)$	I	0.5200	0.3600	0.4700
$p(go\|go)$	I	0.7115	0.7500	0.6383
$p(go\|no\text{-}go)$	I	0.0625	0.2031	0.1887
$p(no\text{-}go\|go)$	I	0.2885	0.2500	0.3617
$p(go)$	II	0.4875	0.4600	0.6333
$p(go\|go)$	II	0.7692	0.5870	0.6053
$p(go\|no\text{-}go)$	II	0.0488	0.2407	0.0455
$p(no\text{-}go\|go)$	II	0.2308	0.4130	0.3947
$p(go)$	III	0.5667	0.4375	0.4800
$p(go\|go)$	III	0.6471	0.6857	0.6875
$p(go\|no\text{-}go)$	III	0.0769	0.1778	0.1346
$p(no\text{-}go\|go)$	III	0.4333	0.3143	0.3125

Table 10.14 Summarization of values from study with operators and time periods.

Operator	p(go)	p(go|go)	p(go|no-go)	p(no-go|go)
A	0.5247	0.7093	0.0627	0.2907
B	0.4192	0.6742	0.2072	0.3258
C	0.5278	0.6437	0.1192	0.3563
Time Period				
I	0.4500	0.6999	0.1514	0.3001
II	0.5269	0.6538	0.1117	0.3462
III	0.4947	0.6734	0.1098	0.3266

Chapter 10 Take-Home Pay

1. Attribute methods can be improved through maximum likelihood fitting.

2. Methods of this type exist for repeatability alone.

3. Methods of this type exist for repeatability and reproducibility.

4. Methods of this type exist for uncertain standards.

5. These methods can be extended to other error sources.

11

Extending the Attribute MSA

T he approach to attribute measurement analysis introduced in Chapter 10 is only the beginning of what can be done to extend this type of analysis to handle more sources of error. These sources can be discrete or continuous in nature, just like in the ANOVA method. Once this extension is made, one can also use it to improve attribute studies when the situation is deformative, destructive, or dynamic. Through the application of this method, one can get as much power and flexibility in an attribute measurement study as one can get in a variables measurement systems study. By including more components of variation in each attribute study, the investigator can obtain a more complete understanding of the measurement effects and compute better estimates of each component.

LOGISTIC ANALYSIS

The secret to achieving these advantages is to apply logistic analysis. There are several ways in which to do this, including generalized linear models (Myers, 2001), but this book will apply logistic regression to the problem at hand. Logistic regression is basically an extension of the ANOVA method to attribute data (Myers, 2001). This neat trick is accomplished by modeling the probability of a good or bad measurement. Unfortunately, this cannot be handled directly by simply substituting the probability p into the regular regression routine. One must work with the transformation log $(p/(1-p))$ instead. With this transformation and a good software package for doing the computations, one can apply logistic regression to all the situations mentioned earlier. Of course, like all procedures, the more one knows about the workings of the method, the better the outcomes will be. Therefore, it will pay to read up on the methods in more detail.

As an example of this logistic regression approach, consider a mocked-up situation in which the probability of a false alarm changes in a linear fashion with true product value. Assume the product value grows from 100 to 110 in increments of one. Also assume that the probability of a measurement error is equal to .01*(true value − 100). Imagine further the collection of 10 random samples from each true part value. Such a set of data might look like that presented in Table 11.1 and also include the computed value of log(p/1 − p).

A plot of this relationship may be easier to assimilate. Figure 11.1 shows a graph of p and log(p/1 − p) versus the true part value. Notice that a

Table 11.1   A listing of probability and its transform versus part value.

Part True Value	p	Log(p/1 − p)
100	0.01	−4.60
101	0.02	−3.89
102	0.03	−3.48
103	0.04	−3.18
104	0.05	−2.94
105	0.06	−2.75
106	0.07	−2.59
107	0.08	−2.44
108	0.09	−2.31
109	0.10	−2.20

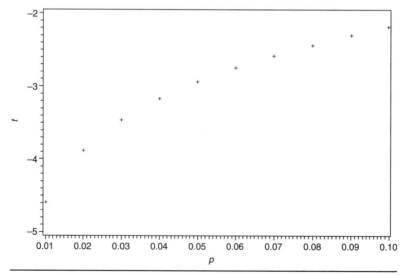

Figure 11.1   The logistic transformation of probability.

linear growth in the probability leads to a nonlinear shape to the $\log(p/1 - p)$ values.

It is perhaps more pertinent to examine what pattern in the probabilities actually does lead to a linear pattern in the transformed values. Table 11.2 shows such a pattern, and Figure 11.2 shows the linear plot.

So the logistic regression finds linear relationships and offsets due to probability changes that conform to these transformation rules. But it is easy enough to transform the values back to the probabilities that generate them.

Table 11.2   A linear relationship of probabilities and transforms.

Part True Value	p	Log(p/1 – p)
100	0.4975	–0.01
101	0.4950	–0.02
102	0.4925	–0.03
103	0.4900	–0.04
104	0.4875	–0.05
105	0.4850	–0.06
106	0.4825	–0.07
107	0.4800	–0.08
108	0.4775	–0.09
109	0.4850	–0.10

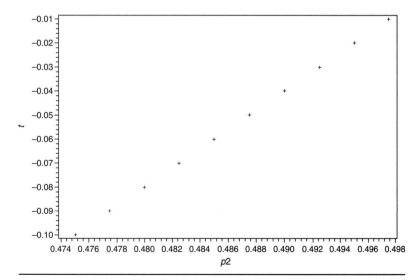

Figure 11.2   Probabilities that lead to linear transformed values.

APPLICATION OF LOGISTIC REGRESSION TO A SIMPLE ATTRIBUTE MSA

Consider the situation of an attribute gage that has only repeatability effects. If the probability of an error does not depend on the true part value, one can collect perhaps 10 parts in which the true classification is known by reference to a standard. Each of the 10 parts is classified by the measurement system in question for 10 repetitions to yield a total of 100 values. For example, such data could be similar to that shown in Table 11.3.

Table 11.3   Attribute MSA data with no operators.

Obs	Parts	Reps	Noncons
1	1	10	0
2	2	10	1
3	3	10	0
4	4	10	0
5	5	10	1
6	6	10	1
7	7	10	0
8	8	10	0
9	9	10	0
10	10	10	1

The logistic regression model contains just an intercept or constant term because the measurement error is assumed not to change in this example. The fitted transformed value is –3.1781, which transforms to a probability value of 0.04. Table 11.4 shows the logistic regression output. Thus the probability of a false positive is estimated to be 0.04 for this measurement system.

Because the false positive and false negative errors cannot occur on the same true product value to estimate the false negative, one must do a separate analysis on data that refer to true nonconforming parts that are potentially misclassified as conforming. Consider another set of 10 parts that are all nonconforming in actuality that are measured 10 repeat times, with the data and results given in Table 11.5. The intercept is –4.5951, which transforms to 0.01 as the probability of a false negative.

Table 11.4 Attribute MSA for false positives without operators.

Parameter	DF	Estimate	Error	Chi-Square	Pr > ChiSq
Intercept	1	–3.1781	0.5103	38.7841	<.0001
	Obs	Parts	Reps	Noncons	Predict
	1	1	10	0	0.04
	2	2	10	1	0.04
	3	3	10	0	0.04
	4	4	10	0	0.04
	5	5	10	1	0.04
	6	6	10	1	0.04
	7	7	10	0	0.04
	8	8	10	0	0.04
	9	9	10	0	0.04
	10	10	10	1	0.04

Table 11.5 Attribute MSA for false negatives without operators.

Parameter	DF	Estimate	Error	Chi-Square	Pr > ChiSq
Intercept	1	–4.5951	1.0050	20.9040	<.0001
	Obs	Parts	Reps	Noncons	Predict
	1	1	10	0	0.01
	2	2	10	0	0.01
	3	3	10	0	0.01
	4	4	10	0	0.01
	5	5	10	1	0.01
	6	6	10	0	0.01
	7	7	10	0	0.01
	8	8	10	0	0.01
	9	9	10	0	0.01
	10	10	10	0	0.01

APPLICATION OF LOGISTIC REGRESSION TO AN RANDR STUDY

Table 11.6 shows a set of data that could be used for an attribute study that seeks to estimate the false positive and false negative error rates for each of the three appraisers. This can be done as a separate study for each set of data (broken up by operator), but there are some advantages to doing the study by estimating an overall false-positive rate and adjusting this up or down for each appraiser's effect. Table 11.7 shows the estimates for the appraisers' effects.

Table 11.6 Attribute MSA data with appraisers.

Obs	Reps	Appraiser	Parts	Noncons	Predict
1	20	1	1	0	0.023614
2	20	1	2	0	0.023614
3	20	1	3	3	0.023614
4	20	1	4	1	0.023614
5	20	1	5	0	0.023614
6	20	1	6	0	0.023614
7	20	1	7	0	0.023614
8	20	1	8	0	0.023614
9	20	1	9	0	0.023614
10	20	1	10	0	0.023614
11	20	1	11	0	0.023614
12	20	1	12	1	0.023614
13	20	1	13	1	0.023614
14	20	1	14	0	0.023614
15	20	1	15	1	0.023614
16	20	1	16	1	0.023614
17	20	1	17	2	0.023614
18	20	1	18	0	0.023614
19	20	1	19	0	0.023614
20	20	1	20	1	0.023614
21	20	1	21	0	0.023614
22	20	1	22	0	0.023614
23	20	1	23	0	0.023614
24	20	1	24	0	0.023614

(continued)

Table 11.6 Attribute MSA data with appraisers.

Obs	Reps	Appraiser	Parts	Noncons	Predict
25	20	1	25	0	0.023614
26	20	1	26	0	0.023614
27	20	1	27	0	0.023614
28	20	1	28	0	0.023614
29	20	1	29	1	0.023614
30	20	1	30	2	0.023614
31	20	1	31	0	0.023614
32	20	1	32	1	0.023614
33	20	1	33	0	0.023614
34	20	1	34	1	0.023614
35	20	1	35	0	0.023614
36	20	1	36	1	0.023614
37	20	2	1	0	0.027778
38	20	2	2	0	0.027778
39	20	2	3	1	0.027778
40	20	2	4	0	0.027778
41	20	2	5	1	0.027778
42	20	2	6	1	0.027778
43	20	2	7	0	0.027778
44	20	2	8	0	0.027778
45	20	2	9	1	0.027778
46	20	2	10	0	0.027778
47	20	2	11	0	0.027778
48	20	2	12	2	0.027778
49	20	2	13	2	0.027778
50	20	2	14	0	0.027778
51	20	2	15	1	0.027778
52	20	2	16	0	0.027778
53	20	2	17	1	0.027778
54	20	2	18	2	0.027778
55	20	2	19	1	0.027778
56	20	2	20	0	0.027778
57	20	2	21	2	0.027778

(continued)

Table 11.6 Attribute MSA data with appraisers.

Obs	Reps	Appraiser	Parts	Noncons	Predict
58	20	2	22	1	0.027778
59	20	2	23	0	0.027778
60	20	2	24	0	0.027778
61	20	2	25	1	0.027778
62	20	2	26	0	0.027778
63	20	2	27	0	0.027778
64	20	2	28	0	0.027778
65	20	2	29	0	0.027778
66	20	2	30	0	0.027778
67	20	2	31	1	0.027778
68	20	2	32	2	0.027778
69	20	2	33	0	0.027778
70	20	2	34	0	0.027778
71	20	2	35	0	0.027778
72	20	2	36	0	0.027778
73	20	3	1	1	0.033336
74	20	3	2	2	0.033336
75	20	3	3	0	0.033336
76	20	3	4	1	0.033336
77	20	3	5	0	0.033336
78	20	3	6	1	0.033336
79	20	3	7	1	0.033336
80	20	3	8	0	0.033336
81	20	3	9	0	0.033336
82	20	3	10	4	0.033336
83	20	3	11	0	0.033336
84	20	3	12	0	0.033336
85	20	3	13	1	0.033336
86	20	3	14	0	0.033336
87	20	3	15	2	0.033336
88	20	3	16	0	0.033336
89	20	3	17	0	0.033336
90	20	3	18	0	0.033336

(continued)

(continued)

Table 11.6 Attribute MSA data with appraisers.

Obs	Reps	Appraiser	Parts	Noncons	Predict
91	20	3	19	2	0.033336
92	20	3	20	0	0.033336
93	20	3	21	3	0.033336
94	20	3	22	1	0.033336
95	20	3	23	1	0.033336
96	20	3	24	0	0.033336
97	20	3	25	1	0.033336
98	20	3	26	0	0.033336
99	20	3	27	0	0.033336
100	20	3	28	0	0.033336
101	20	3	29	0	0.033336
102	20	3	30	0	0.033336
103	20	3	31	0	0.033336
104	20	3	32	0	0.033336
105	20	3	33	0	0.033336
106	20	3	34	1	0.033336
107	20	3	35	1	0.033336
108	20	3	36	1	0.033336

Table 11.7 An MSA with appraisers.

Parameter	DF	Estimate	Error	Chi-Square	Pr > ChiSq
Intercept	1	−3.5482	0.1311	732.1100	<.0001
Appraiser 1	1	−0.1738	0.1931	0.8107	0.3679
Appraiser 2	1	−0.00715	0.1853	0.0015	0.9692
Appraiser 3	—	—	—	—	−(−0.1738 − 0.00715) = 0.18095

Here is the output from the logistic regression with appraiser as a discrete input.

Intercept = –3.5482   Appraiser 1 offset = –0.1738
Appraiser 2 offset = –0.00715 Appraiser 3 offset = 0.18095

To use this to estimate the probability of false positive for appraiser 1, one simply adds the intercept + the appraiser 1 offset to get –3.5482 – 0.1738 = –3.722. The transformed value is .0236; for appraiser 1 it is 0.0278, and for appraiser 3 it is .0333. Table 11.8 shows another set of data for the false negative, and the results are given in Table 11.9.

Table 11.8 Attribute MSA for false negative.

Obs	Reps	Appraiser	Parts	Noncons	Predict
1	20	1	1	0	0.023612
2	20	1	2	0	0.023612
3	20	1	3	3	0.023612
4	20	1	4	1	0.023612
5	20	1	5	0	0.023612
6	20	1	6	0	0.023612
7	20	1	7	0	0.023612
8	20	1	8	0	0.023612
9	20	1	9	0	0.023612
10	20	1	10	0	0.023612
11	20	1	11	0	0.023612
12	20	1	12	1	0.023612
13	20	1	13	1	0.023612
14	20	1	14	0	0.023612
15	20	1	15	1	0.023612
16	20	1	16	1	0.023612
17	20	1	17	2	0.023612
18	20	1	18	0	0.023612
19	20	1	19	0	0.023612
20	20	1	20	1	0.023612
21	20	1	21	0	0.023612
22	20	1	22	0	0.023612

(continued)

Table 11.8 Attribute MSA for false negative.

Obs	Reps	Appraiser	Parts	Noncons	Predict
23	20	1	23	0	0.023612
24	20	1	24	0	0.023612
25	20	1	25	0	0.023612
26	20	1	26	0	0.023612
27	20	1	27	0	0.023612
28	20	1	28	0	0.023612
29	20	1	29	1	0.023612
30	20	1	30	2	0.023612
31	20	1	31	0	0.023612
32	20	1	32	1	0.023612
33	20	1	33	0	0.023612
34	20	1	34	1	0.023612
35	20	1	35	0	0.023612
36	20	1	36	1	0.023612
37	20	2	1	1	0.037500
38	20	2	2	0	0.037500
39	20	2	3	1	0.037500
40	20	2	4	0	0.037500
41	20	2	5	1	0.037500
42	20	2	6	2	0.037500
43	20	2	7	0	0.037500
44	20	2	8	1	0.037500
45	20	2	9	1	0.037500
46	20	2	10	0	0.037500
47	20	2	11	0	0.037500
48	20	2	12	2	0.037500
49	20	2	13	2	0.037500
50	20	2	14	1	0.037500
51	20	2	15	1	0.037500
52	20	2	16	1	0.037500
53	20	2	17	1	0.037500
54	20	2	18	3	0.037500
55	20	2	19	1	0.037500

(continued)

Table 11.8 Attribute MSA for false negative.

Obs	Reps	Appraiser	Parts	Noncons	Predict
56	20	2	20	0	0.037500
57	20	2	21	2	0.037500
58	20	2	22	1	0.037500
59	20	2	23	0	0.037500
60	20	2	24	0	0.037500
61	20	2	25	1	0.037500
62	20	2	26	0	0.037500
63	20	2	27	0	0.037500
64	20	2	28	0	0.037500
65	20	2	29	0	0.037500
66	20	2	30	0	0.037500
67	20	2	31	1	0.037500
68	20	2	32	3	0.037500
69	20	2	33	0	0.037500
70	20	2	34	0	0.037500
71	20	2	35	0	0.037500
72	20	2	36	0	0.037500
73	20	3	1	2	0.058334
74	20	3	2	3	0.058334
75	20	3	3	1	0.058334
76	20	3	4	2	0.058334
77	20	3	5	0	0.058334
78	20	3	6	1	0.058334
79	20	3	7	1	0.058334
80	20	3	8	0	0.058334
81	20	3	9	0	0.058334
82	20	3	10	5	0.058334
83	20	3	11	0	0.058334
84	20	3	12	0	0.058334
85	20	3	13	1	0.058334
86	20	3	14	0	0.058334
87	20	3	15	2	0.058334
88	20	3	16	1	0.058334

(continued)

(continued)

Table 11.8 Attribute MSA for false negative.

Obs	Reps	Appraiser	Parts	Noncons	Predict
89	20	3	17	0	0.058334
90	20	3	18	1	0.058334
91	20	3	19	2	0.058334
92	20	3	20	1	0.058334
93	20	3	21	4	0.058334
94	20	3	22	2	0.058334
95	20	3	23	1	0.058334
96	20	3	24	1	0.058334
97	20	3	25	1	0.058334
98	20	3	26	1	0.058334
99	20	3	27	0	0.058334
100	20	3	28	0	0.058334
101	20	3	29	0	0.058334
102	20	3	30	0	0.058334
103	20	3	31	1	0.058334
104	20	3	32	1	0.058334
105	20	3	33	1	0.058334
106	20	3	34	2	0.058334
107	20	3	35	2	0.058334
108	20	3	36	2	0.058334

Table 11.9 Attribute analysis for false negative.

Parameter	DF	Estimate	Error	Chi-Square	Pr > ChiSq
Intercept	1	−3.2496	0.1174	766.3852	<.0001
Appraiser 1	1	−0.4725	0.1840	6.5942	0.0102
Appraiser 2	1	0.00440	0.1631	0.0007	0.9785
Appraiser 3	—	—	—	—	−(−0.4725 + .00440) = 0.4685

EXTENSION OF THE MSA TO OTHER DISCRETE SOURCES OF DISPERSION

Here is a set of data for a four-source MSA study. The four sources are appraisers, parts, setups, and machines. Repeatability is estimated as a residual error, just as in the ANOVA components approach for a variable MSA. Table 11.10 shows the data, and Table 11.11 shows the output of the logistic regression with the computed probabilities of false positive and false negative for each combination.

Table 11.10 Attribute data with more sources.

Index	Reps	Parts	Appraiser	Machines	Setups	Noncons	Predict
1	4	1	1	1	1	0	0.04450
2	4	1	1	1	2	0	0.07947
3	4	1	1	2	1	1	0.03663
4	4	1	1	2	2	1	0.06584
5	4	1	2	1	1	0	0.05397
6	4	1	2	1	2	0	0.09563
7	4	1	2	2	1	0	0.04450
8	4	1	2	2	2	0	0.07947
9	4	2	1	1	1	0	0.06762
10	4	2	1	1	2	0	0.11850
11	4	2	1	2	1	0	0.05590
12	4	2	1	2	2	1	0.09890
13	4	2	2	1	1	1	0.08158
14	4	2	2	1	2	0	0.14138
15	4	2	2	2	1	0	0.06762
16	4	2	2	2	2	1	0.11850
17	4	3	1	1	1	1	0.04450
18	4	3	1	1	2	1	0.07947
19	4	3	1	2	1	0	0.03663
20	4	3	1	2	2	0	0.06584
21	4	3	2	1	1	0	0.05397
22	4	3	2	1	2	0	0.09563

(continued)

(continued)

Table 11.10 Attribute data with more sources.

Index	Reps	Parts	Appraiser	Machines	Setups	Noncons	Predict
23	4	3	2	2	1	0	0.04450
24	4	3	2	2	2	0	0.07947
25	4	4	1	1	1	0	0.09133
26	4	4	1	1	2	0	0.15705
27	4	4	1	2	1	0	0.07583
28	4	4	1	2	2	0	0.13202
29	4	4	2	1	1	1	0.10961
30	4	4	2	1	2	2	0.18580
31	4	4	2	2	1	0	0.09133
32	4	4	2	2	2	1	0.15705

Table 11.11 Attribute analysis for more sources.

Parameter	DF	Estimate	Error	Chi-Square	Pr > ChiSq
Intercept	1	−2.4551	0.3415	51.6986	<.0001
Appraiser 1	1	−0.1014	0.3193	0.1009	0.7507
Part 1	1	−0.3030	0.6139	0.2437	0.6216
Part 2	1	0.1399	0.5423	0.0665	0.7965
Part 3	1	−0.3030	0.6139	0.2437	0.6216
Machine 1	1	−0.3086	0.3284	0.8830	0.3474
Setup 1	1	0.1014	0.3193	0.1009	0.7507

EXTENSION OF THE MSA TO CONTINUOUS SOURCES OF DISPERSION

Now consider an MSA in which a continuous input is available in addition to repeatability. Table 11.12 shows the data, and tables 11.13 and 11.14 show the output and computed probabilities for several settings of temperature.

Table 11.12   Attribute data for continuous sources.

Obs	Reps	Temp	Noncons	Predict
1	10	80	0	0.02602
2	10	81	0	0.02863
3	10	82	2	0.03149
4	10	83	1	0.03463
5	10	84	0	0.03806
6	10	85	0	0.04183
7	10	86	0	0.04594
8	10	87	0	0.05044
9	10	88	0	0.05536
10	10	89	0	0.06072
11	10	90	0	0.06657
12	10	91	2	0.07293
13	10	92	1	0.07985
14	10	93	0	0.08737
15	10	94	1	0.09552
16	10	95	2	0.10434
17	10	96	2	0.11388
18	10	97	1	0.12417
19	10	98	1	0.13524
20	10	99	2	0.14714
21	10	100	1	0.15989

Table 11.13   Attribute analysis for continuous source.

Parameter	DF	Estimate	Error	Chi-Square	Pr > ChiSq
Intercept	1	−11.4755	4.3883	6.8384	0.0089
Temp	1	0.0982	0.0472	4.3325	0.0374

Table 11.14 Prediction for the continuous source.

Obs	Reps	Temp	Noncons	Predict
1	10	80	0	0.076190
2	10	81	0	0.076190
3	10	82	2	0.076190
4	10	83	1	0.076190
5	10	84	0	0.076190
6	10	85	0	0.076190
7	10	86	0	0.076190
8	10	87	0	0.076190
9	10	88	0	0.076190
10	10	89	0	0.076190
11	10	90	0	0.076190
12	10	91	2	0.076190
13	10	92	1	0.076190
14	10	93	0	0.076190
15	10	94	1	0.076190
16	10	95	2	0.076190
17	10	96	2	0.076190
18	10	97	1	0.076190
19	10	98	1	0.076190
20	10	99	2	0.076190
21	10	100	1	0.076190

EXTENSION TO DESTRUCTIVE ATTRIBUTE MSA

Consider an attribute study in which a series of parts is presented for measurement, but there is a hidden effect that changes the true part value so that the probability of false positive increases in each repetition. Assuming that the change is linear, one can add a factor to the logistic regression that marks the order of the measurement. Table 11.15 shows the example data, and tables 11.16 through 11.18 show the analysis results and the associated probabilities.

Table 11.15 Data for a destructive attribute MSA.

Obs	Reps	Order	Noncons	Predict
1	1	1	0	0.10046
2	1	2	0	0.09900
3	1	3	1	0.09755
4	1	4	0	0.09612
5	1	5	0	0.09472
6	1	6	0	0.09333
7	1	7	0	0.09195
8	1	8	0	0.09060
9	1	9	0	0.08927
10	1	10	0	0.08795
11	1	11	0	0.08665
12	1	12	0	0.08537
13	1	13	0	0.08410
14	1	14	0	0.08285
15	1	15	0	0.08162
16	1	16	0	0.08041
17	1	17	1	0.07921
18	1	18	0	0.07803
19	1	19	0	0.07686
20	1	20	0	0.07571
21	1	21	0	0.07458
22	1	22	0	0.07346
23	1	23	0	0.07236
24	1	24	0	0.07127
25	1	25	0	0.07020
26	1	26	0	0.06914
27	1	27	0	0.06810
28	1	28	0	0.06707
29	1	29	0	0.06605
30	1	30	1	0.06505
31	1	31	0	0.06407
32	1	32	0	0.06310
33	1	33	0	0.06214
34	1	34	0	0.06120

(continued)

(continued)

Table 11.15 Data for a destructive attribute MSA.

Obs	Reps	Order	Noncons	Predict
35	1	35	0	0.06027
36	1	36	0	0.05935
37	1	37	0	0.05844
38	1	38	0	0.05755
39	1	39	0	0.05667
40	1	40	0	0.05581
41	1	41	0	0.05495
42	1	42	0	0.05411
43	1	43	0	0.05328
44	1	44	0	0.05247
45	1	45	0	0.05166
46	1	46	0	0.05087
47	1	47	0	0.05009
48	1	48	0	0.04932
49	1	49	0	0.04856
50	1	50	0	0.04781
51	1	51	0	0.04707
52	1	52	0	0.04634
53	1	53	0	0.045629
54	1	54	1	0.044924
55	1	55	0	0.044229
56	1	56	0	0.043545
57	1	57	0	0.042870
58	1	58	0	0.042206
59	1	59	0	0.041551
60	1	60	0	0.040906

Table 11.16 Analysis of the destructive attribute MSA.

Parameter	DF	Estimate	Error	Chi-Square	Pr > ChiSq
Intercept	1	−2.1758	0.9570	5.1694	0.0230
Order	1	−0.0163	0.0306	0.2851	0.5934

Table 11.17 The effect of the continuous source.

Parameter	DF	Estimate	Error	Chi-Square	Pr > ChiSq
Intercept	1	−2.6391	0.5175	26.0013	<.0001

Table 11.18 The prediction for the destructive attribute MSA.

Obs	Reps	Order	Noncons	Predict
1	1	1	0	0.066667
2	1	2	0	0.066667
3	1	3	1	0.066667
4	1	4	0	0.066667
5	1	5	0	0.066667
6	1	6	0	0.066667
7	1	7	0	0.066667
8	1	8	0	0.066667
9	1	9	0	0.066667
10	1	10	0	0.066667
11	1	11	0	0.066667
12	1	12	0	0.066667
13	1	13	0	0.066667
14	1	14	0	0.066667
15	1	15	0	0.066667
16	1	16	0	0.066667
17	1	17	1	0.066667
18	1	18	0	0.066667
19	1	19	0	0.066667
20	1	20	0	0.066667
21	1	21	0	0.066667
22	1	22	0	0.066667
23	1	23	0	0.066667
24	1	24	0	0.066667
25	1	25	0	0.066667
26	1	26	0	0.066667
27	1	27	0	0.066667

(continued)

(continued)

Table 11.18 The prediction for the destructive attribute MSA.

Obs	Reps	Order	Noncons	Predict
28	1	28	0	0.066667
29	1	29	0	0.066667
30	1	30	1	0.066667
31	1	31	0	0.066667
32	1	32	0	0.066667
33	1	33	0	0.066667
34	1	34	0	0.066667
35	1	35	0	0.066667
36	1	36	0	0.066667
37	1	37	0	0.066667
38	1	38	0	0.066667
39	1	39	0	0.066667
40	1	40	0	0.066667
41	1	41	0	0.066667
42	1	42	0	0.066667
43	1	43	0	0.066667
44	1	44	0	0.066667
45	1	45	0	0.066667
46	1	46	0	0.066667
47	1	47	0	0.066667
48	1	48	0	0.066667
49	1	49	0	0.066667
50	1	50	0	0.066667
51	1	51	0	0.066667
52	1	52	0	0.066667
53	1	53	0	0.066667
54	1	54	1	0.066667
55	1	55	0	0.066667
56	1	56	0	0.066667
57	1	57	0	0.066667
58	1	58	0	0.066667
59	1	59	0	0.066667
60	1	60	0	0.066667

Chapter 11 Take-Home Pay

1. Logistic regression can provide a general approach for attribute MSA.

2. Logistic regression can handle the standard RandR analysis.

3. Logistic regression can be extended to multiple error sources.

4. Logistic regression can provide a method to handle continuous error sources.

5. Logistic regression can help with destructive, deformative, and dynamic MSA studies.

12

The Evaluation of Measurement System Value

C hapter 1 surveyed many of the potential impacts of measurement error on product classification, process control, and knowledge management. When these errors occur, some cost is added to the process. Sometimes this cost can be small, and often it is hidden. But it is there, and its reduction or elimination must be worth something. This chapter seeks to introduce methods by which one can assign an explicit cost or risk to the presence of measurement error. It also seeks to show how to utilize these risk methods to manage an effective measurement system management program so that the bottom line benefits.

THE CONCEPT OF RISK

The unifying principle behind this assessment approach is the idea of risk. For the purposes of this chapter, *risk* is defined as the product of the loss attributable to an error multiplied by the probability of that error (Henley, 1981). For example, the potential loss attributable to nuclear power plant release might easily be estimated as a billion dollars, but if the probability of a mishap is controlled to less than one chance in a million, the risk is one billion dollars times a probability of one millionth, which is equal to one thousand dollars. This is a definite value that can then be evaluated along with risks due to coal plant accidents and other mishaps. Notice that because both the loss and probability values depend on other factors, such as duration of the event, one must be very careful to define the limitations and scope of the risk that is computed.

RISK ON PRODUCT SORTING

Consider this risk approach for a single measurement sorting operation. That is, the product is measured one time by a system with known (or estimated) measurement error. If this single measurement exceeds an engineering specification, it is scrapped. If the single measurement nestles inside the specifications, it is passed on to the customer with no further checks made on it. For example, assume the measurement system has an overall uncertainty standard deviation of 10 mm with zero bias. This implies that the measurement errors can be thought of as random choices from a normal distribution with standard deviation = 10 mm and mean = 0 mm. One of these random errors is added to each and every true part value to produce a unique measured value. It is this measured value as the sum of true value plus measurement error that is used to classify the part and, ultimately, to determine its cost to the process. Table 12.1 shows a set of such measurements that could result from measuring a sequence of 10 parts each of true part value 100 mm.

If the tolerances are 0 mm and 200 mm, one can intuitively see that there is little chance that measurement errors alone will result in any scrapped product. But if the tolerances are 90 to 110 mm, there is a serious potential impact. This analysis is missing out on the fact that the true part values may also vary. After all, they are coming from a production process that is not likely to produce precisely the same value again and again. If the true (without measurement error) process capability is 1, the true parts distribution will just fit into the tolerance limits (Ryan, 1989), as shown in Figure 12.1.

If measurement error is added to these true values, the distribution of measured values will look like Figure 12.2. Notice that the shape is the same between the two figures, but the second one is wider. Note that even though the measurement errors average zero and are merely pluses and minuses from each true value, the variation of the combined population gets larger. This is the case for almost all measurement systems.

Table 12.1 A simulation of a distribution of measurements.

Obs	Part	Truepart	Measerr	Measpart
1	1	100	13.3916	113.392
2	2	100	2.2365	102.236
3	3	100	−10.6236	89.376
4	4	100	5.7616	105.762
5	5	100	−4.5478	95.452
6	6	100	−10.7551	89.245
7	7	100	14.5509	114.551
8	8	100	20.5337	120.534
9	9	100	−13.8463	86.154
10	10	100	7.3842	107.384

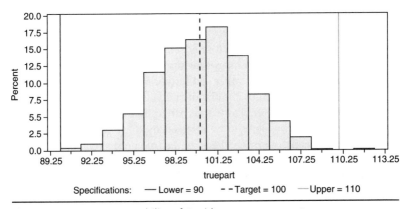

Figure 12.1 Process capability of 1 with no measurement error.

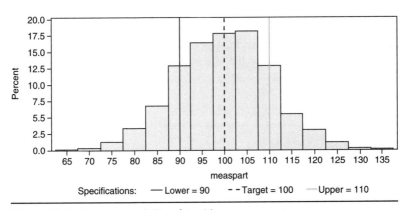

Figure 12.2 Process capability of 1 with measurement error.

THE COST COMPONENT

To compute risk, one also needs to know the loss or cost incurred when a mistake is made. There are two kinds of mistakes that can be made in the product sorting situation. First, a part with a true value inside the specifications can measure outside and be scrapped. Second, a part with a true value outside the specifications can measure inside and be shipped to the customer. There is some cost for scrapping, which is equal to the manufacturing cost of the part. Assume this is $1 as an example. There is also a cost incurred when the customer uses the out-of-specification part. This cost can vary widely depending on the situation, but it is almost always larger than the scrapping cost: Assume $100 as an example. The risk is the combination of these losses and the probabilities of the errors caused by the measurement system error. In this example, the loss is constant regardless of the precise value of the part value. But in many real circumstances, the cost will grow greater as the part value gets further and further from the target for the process. However, the probability of a classification decision error does depend on the true value of the part. Those true part values that are nearest to the tolerances are the most likely to be affected by these errors. Figure 12.3 shows the relationship of error and cost to the true product value.

The summary statistics of these costs are shown in Table 12.2.

Often the average or expected value of this curve is used as a summary of the risk introduced by the measurement system error. Notice that this risk depends on the true value distribution as well. Table 12.2 is a demonstration of risk for a process with true capability of 1.00, whereas Table 12.3 shows the risk summary for a process with true capability of 1.33.

Notice that product sorting risk can be reduced either by improving measurement performance or by improving true process performance. Table 12.4 shows the summary statistics for a capability 1.00 process with a measurement system that has half the original error.

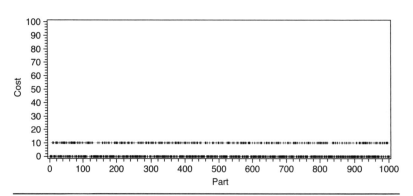

Figure 12.3 A sequence of measurements with their cost.

Table 12.2 Summary statistics of misclassification cost.

N	Mean	Std Dev	Minimum	Maximum
1000	3.5600000	5.6531724	0	100.0000000

Table 12.3 Statistics of cost for a better process.

N	Mean	Std Dev	Minimum	Maximum
1000	3.2700000	4.6935199	0	10.0000000

Table 12.4 Statistics of costs for a barely capable process.

N	Mean	Std Dev	Minimum	Maximum
1000	0.9700000	4.2162634	0	100.0000000

RISK EVALUATION FOR ATTRIBUTE GAGES

The risk approach can also be useful in the evaluation of attribute gages. If the attribute gage analysis is a simple one, it results in two simple probabilities: the probability of a false negative, $p(\text{fn})$, and the probability of a false positive, $p(\text{fp})$. These probabilities by definition vary only in the simplest way with the true part value. Essentially these errors are zero when the decision is correct and equal to $p(\text{fn})$ or $p(\text{fp})$ otherwise.

If the costs are simple, as in the variable gage case, the probabilities and the costs for a typical attribute gage might be similar to those shown in Table 12.5. To compute risk, one needs to include information on the incidence rate of true conforms and nonconforms (only one is necessary because they must add up to one). The average risk is simply the product of $p(\text{true conforms})*p(\text{nonconforming given conforming})*\text{cost(false nonconform)} + p(\text{true nonconform})*p(\text{conforming given nonconform})*\text{cost(false conform)}$. For example, using the probabilities and costs from Table 12.5

Table 12.5 Cost calculations for an attribute gage.

Event	Probability	Cost	Risk
Measured C True NC	0.0020 = 0.10 × 0.02	$100	$0.20
Measured NC True C	0.0450 = 0.90 × 0.05	$10	$0.45
Measured C True C	0.855 = 0.90 × 0.95	$0	$0.00
Measured NC True NC	0.0980 = 0.10 × 0.98	$10	$0.98
Total			$1.63

and assuming that true nonconforms occur with a frequency of 1 percent, the average risk is $1.63.

One use of the risk analysis is the comparison of different scenarios. One such scenario is a change in the true product process. If the probability of a nonconform is 5 percent instead of 1 percent, the average risk is $1.065. Another interesting scenario is a change in the measurement system. Perhaps the probability(false alarm) and the probability(true alarm) are reduced to half their original value. The expected risk assumed under this new scenario is $1.315.

RISK ASSESSMENT

Risk as defined here requires understanding of the measurement system performance, the process performance, and the costs associated with the various decision errors. Therefore, in a real application one must identify and maintain the process capability, the measurement capability, and the costs. Although all these activities require work to obtain the data, it is surprising that often it is the cost information that is the most difficult to obtain. For example, if the cost of a false positive is to be assessed, one has to consider several things. First of all, there is the issue of which costs to include in the calculation. Some of the potential costs associated with the false positive are inspection costs, transfer costs, production time loss, scrap processing time, scrap processing resource, lost material, and maintenance costs. Second, along with the type of costs, there is the issue of what horizon to apply. For example, the false-positive product may sit in a scrap storage area awaiting disposal for a long time. Does a product stop accumulating costs after one year? After one day? Or are costs instantaneous? A third issue relates to the time value of money. Is the inspection cost of a particular measurement error a sunk cost or is it variable? Should it be depreciated? Obviously there are a multitude of issues related to cost, if one wants an exact accounting value of the impact of measurement error.

Practically, the best way to proceed is to accept the definitions that the accounting system of the company requires. For projects there is certain to be an evaluation method that is to be applied with all the important choices already defined. This cost evaluation method is not absolutely complete, but it is likely to be exact. So it should serve as a de facto standard for comparing alternate measurement systems and for process improvements. The use of this risk method will be even easier to apply in those companies and situations in which risk is already a part of the evaluation process. Perhaps the only two shortcomings of these standardized evaluation systems for measurement system evaluation are that they tend to be short term (on the order of one to five years) and they are not adept at capturing hidden costs. However, measurement systems are applied so often that even small error rates can have large impacts on the bottom line.

APPLYING MEASUREMENT RISK PROGRAMS

The purpose of a measurement risk management program is to control and reduce the risks of the use of measurement systems in an organization (Henley, 1981). Although this may seem strange at first, the concept is not very different from the management of inspection programs or research and development programs. There are at least four important approaches that can be valuable in this measurement management effort. First, one can decide where and when to employ the measurement systems. A destructive measurement, for example, must be applied sparingly if it is to be cost- or risk-effective. Similarly, any measurement has a cost and associated value that must be balanced. Second, one can decide which components of variation to estimate. If there is a single dominant source of error, it may not be cost-justified to track all the other lesser components. Third, one can decide how well to estimate the various components. Presumably it costs less to estimate just a few sources of error compared with a full-blown study of the entire list of uncertainties associated with a device. It makes sense in many situations to allow poorer estimations for small components than for the larger ones. Fourth, one must choose an action in response to information about the measurement system. These actions could include the following: (1) improve the measurement system, (2) stop using the system, (3) change the frequency or intensity of application of the measurement, (4) alter the way the system is used, (5) improve the production system performance, (6) make the decision process robust to the measurement errors, (7) alter the relationship to cost, and (8) any combination of these actions.

The first approach deals with choices about where to take the measurement and when to take it. Often these two questions are the same for production processes in which the manufacturing order is sequential. But even in these cases there are often choices to be made that can affect the risk performance. For example, consider a measurement that can be taken immediately after an extrusion process or that can be taken later in the process after cooling has occurred. The temperature of the product probably affects the measurement error. So any differences in timing can lead directly to differences in measurement errors. One choice that can be made if the measurement system is too variable is to delay it until the product has cooled. On the other hand, if the measurement is used to adjust the extrusion process, a delay in the feedback may lead to long runs of off-target product characteristics or other control problems. The placement or timing of the measurement can affect both the probability of an error and the cost that is associated with a particular error.

The second approach to measurement risk management involves the choice of how many components of error to study or track. The standard RandR approach basically makes the decision for the investigator: the only two important sources of error to identify are repeatability and reproducibility. Usually it is recommended to reestimate these same two sources whenever events occur that could change them significantly. Such events could include large changes in the design of either the measurement system or the production system. Personnel changes where adequate training is neglected should also be of concern. This approach makes good sense for many measurement systems, but clearly it does not cover all situations. If setup is more important than appraiser, it should replace reproducibility in the most-tracked list.

Another approach that is useful when deciding how many components to track is to start with a complete study that isolates and estimates all the components of uncertainty that are likely to be large. Then, for each of these components, one can do a failure modes and effects analysis (FMEA) to evaluate how often each may change (Stamatis, 2003). That is, how sensitive is this component likely to be to systems changes? For example, an appraiser work method may change only if there is significant turnover in the pool of operators. The effect of temperature may be important only at the extremes of summer and winter, and so on. A plan can be put together that can track the different components at different frequencies according to their sensitivities to change and the sensitivity of the decision to them.

The third approach concerns the quality of the estimation that is done on each component. A measurement error that is 10 times smaller than the dominant one may still cause losses, but it is likely to cause much fewer losses. If the FMEA-style analysis also includes the cost or impact of the error caused by each component, one can use this information to judge the value of tracking the component. Some kind of schedule can then be made that justifies the extra expense of tracking more sources in terms of the extra value it gives to the product itself.

The fourth approach deals with the type of responses or actions that one can take on the basis of the measurement system analyses or tracking. This action can be quite varied in intensity, duration, and success. Cost must be used to direct the choices between these actions. Rather than detail every possible action, the discussion will center on a few of the more unusual approaches.

RISK IMPROVEMENT APPROACHES

Sampling plans can be applied to the measurement system management system (Schilling, 1982). A small set of devices from a pool of them may be evaluated periodically, say every week, with a small study. If some of the sampled devices fail, other larger sets may be taken. Assuming that poorly performing devices are corrected, the overall quality of the measurement system pool can be maintained to a given level. For example, one may be agreeable to having an average of no more than 5 percent of all devices performing out of specification at any given time. This can be done with either an attribute or a variable acceptance sampling plan.

If a single measurement is poor, one might still be able to use the same system if multiple readings can be substituted for a single one. For example, a device with a repeatability error of 50 percent could easily improve to 25 percent simply by taking four independent measurements of the same product and basing product disposition on the average of the four values. Both repeatability and reproducibility might be improved by having several appraisers make the repeat measurements or by using several different measurement instruments. Of course, such repeated measurements are likely to cost more than a single one, but this is part of the risk trade-off strategy that is to be developed.

One must be careful in applying the idea of multiple measurements in a halfway fashion that is misused in many real applications. Often if the first single measurement results in a product measurement that is out of tolerance, a second reading is taken. If the second measurement replaces the first, the piece will be rejected only if the measurement distorts the true value in roughly the same way twice in a row. In most cases this new probability is equal to the probability of a false positive squared. Note that an original value that lies within the tolerances does not usually receive a second reading, so the probability of a false negative is not affected even though it is usually the most costly error. If one's rules allow further repeat measurements of failures, the system is even more distorted.

Another perhaps more novel approach is to try to make the systems more robust to changes. Along the lines that Taguchi has advocated, one may seek to understand what settings of the systems reduce the inherent sensitivity to noise. One way in which to accomplish this is to set up a robust parameter design with the known components of uncertainty fully exercised in the study. The analysis of such an experiment will provide settings (if they exist) for the production system that make it most stable in face of the errors. Or one could choose settings of the measurement system itself that make it more robust to those environmental factors that cause more measurement error.

A last suggestion is to change the relationship of incurred costs to the effect of measurement error. If this is a theoretical relationship enshrined in the company's accounting system, one can seek to change the rules through administrative efforts. If the costs are measured directly on the systems, as would be the case for yield percentages or processing times, one can use something similar to the Taguchi approach but using variation in cost rather than absolute variation. In this way, one seeks to find process settings that minimize or stabilize the systems against possible measurement error influences.

MAINTENANCE OF A MEASUREMENT MANAGEMENT PROGRAM

It is also critical that a measurement risk management program fit into the company's way of doing business. Pilot programs often fail because they are not accepted into the company's overall philosophy, not for technical reasons. To achieve this integration, the values must align with company goals and be evaluated in the same way as more concrete investments, such as equipment upgrades and new pricing. One common difficulty with technically oriented improvement programs is that management is expeted to do the extra work to appreciate the details of the method. Although some companies may be so infused with Six Sigma training that all managers are amateur statisticians, this is not likely to be the case for the vast majority of those who want to understand their measurement systems. Remember, it is up to the investigator to align the actions and results with the prevailing quality programs of the company.

Another important ingredient is to follow up on results. Often there is a placebo-like effect, even for projects. The system improvement may be obvious and dramatic the first few days after implementation, but real success must be long-lived. A life-cycle approach to tracking is probably the ideal way in which to do this kind of thing, but management must already understand this approach or a lot of wheel-spinning will result. A last bit of advice is to remember that the visibility and consistency of effort often matter more than sheer ingenuity or gain. The need for persistence in leading the program and in selling its virtues over and over cannot be underestimated if success is the ultimate target.

Chapter 12 Take-Home Pay

1. Reducing measurement error has measurable value to a company.

2. Risk is the expected loss due to measurement error.

3. Risk can be computed for variable gages and attribute gages.

4. Measurement value should be managed as a program.

5. A measurement value program must be maintained.

13

The Reliability of Measurement Systems

THE DYNAMIC NATURE OF MEASUREMENT SYSTEMS

A measurement system contains physical components. These components might be the working parts of the device or even the people who serve as appraisers in the system. For example, a digital micrometer has mechanical jaws, threads, and electronics that are required to produce the final readout. An operator also uses a micrometer in some specific measurement environments. More complicated systems might integrate many different mechanical, electrical, and personnel systems. One certainty is that all of these physical components will change with time and use. If the components of a measurement system change with time or use, the measurement error can be increased. It is therefore important for measurement systems management to understand and deal with these changes.

It is possible to study the change of performance over time of devices and systems through the use of reliability techniques. *Reliability* is the probability that a device or system will be available for performance at a particular time (Nelson, 1982). It is by nature a statistical quantity and usually must be estimated from data. For measurement systems, one can study the degradation of performance through reliability methods that analyze the failure rates of these systems. Many measurement devices can be corrected and returned to service as recurrent devices (Lawless, 1995), but this chapter will show the reliability approach only for complete failures.

FAILURE-BASED RELIABILITY FOR MEASUREMENT SYSTEMS

A complete measurement system may completely fail, such as when the electronic components of a fluidimeter fail, or there may be a failure of a component part, such as when a piece breaks off the device or plastic fatigue sets in. Even the human components of a measuring system may fail when an accident incapacitates them or they neglect to pay attention to their task. Again, these events can be treated as complete failures of the system.

Many measurement systems merely degrade their performance with time or repeated use. This degradation can be affected by temperature, pressure, current fluctuations, and many other variables. A common form of degradation is caused by friction from the repeated use of mechanical parts in the device. Also, lubrication buildup or material deformation can lead to this kind of degraded performance. If one can set a threshold of performance for a measurement system beyond which the device is no longer considered acceptable, this can also be considered as a complete failure of the system. Of course, this type of defined failure may be more difficult to detect in a system than one of the more complete failures listed earlier. If such a threshold cannot be set or if one does not wish to do so, one may still study the measurement system reliability through the application of degradation analysis methods. Figure 13.1 illustrates a possible relationship between the amount of usage and the reliability of a measurement system.

Failure Modes

There are at least four important features that must be considered in the analysis of complete failures of measurement systems. First, there is the

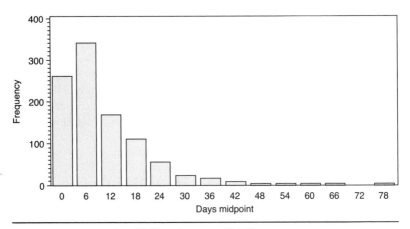

Figure 13.1 Example of failure times as reliability.

idea of a failure mode. A device or system such as a digital micrometer may completely fail for a variety of physical causes. For example, the mechanical jaws may fail after extensive use in a consistent way. If the failure times associated with this component are from a single statistical distribution, it can be considered as a single failure mode. The electronics of the same micrometer may have several failure modes. Perhaps there is a failure due to excessive vibration that is characteristically different from another failure caused by voltage fluctuations. If each of these failure types can be best characterized by separate statistical distributions, they should be considered different failure modes. Because the statistical analysis of reliability is usually pursued for each failure mode separately, it becomes important to clearly define the precise failure type.

Parametric Analysis

Another important feature of reliability analysis of complete failure measurement systems is the use of parametric or nonparametric methods. *Nonparametric methods* make only very generic assumptions about the statistical distribution that characterizes each failure mode (Allison, 1995). They only assume things like smooth and increasing distributions, whereas *parametric methods* assume a single-family or class of distributions. For example, common distributions used in parametric reliability analysis include the exponential, the lognormal, and the Weibull. It is the Weibull distribution that is often used and that will be the choice for examples in this chapter. Once a class of distributions is chosen, the task for the analyst becomes one of finding or estimating the distribution's parameters, hence the name *parametric analysis*. Usually the parametric analysis allows one to compute closer, finer estimates of reliability than the nonparametric methods. However, wrong choices of the distribution can make the parametric analysis less robust than the nonparametric one.

Censoring Methods

A third ingredient of a successful reliability approach is the use of censored data (Nelson, 1982). Systems often have very long times to failure that cannot be tolerated for economic reasons. If the actual failure time of a measurement system is replaced by a time at which the system was removed from the test for another reason, this is termed a *censored value*. Another reason for censoring includes failure of the system for a mode other than the one under study. Reliability methods allow one to make use of the censored data in conjunction with the actual failure times to give better estimates of the distribution's parameters. This also has huge positive implications for the cost of reliability studies.

Variable Wear-Out Rates

Finally, the reliability of a measurement system may be a function of other variables besides wear and stress. In the case of the digital micrometer's mechanical components that deteriorate with wear, it may be the case that low humidity in the workshop accelerates the frictional effects. Temperature may also soften the metal and make it more susceptible to wear out. If the failure modes can be induced in a quicker predictable fashion by adjusting some background variables, one can run accelerated life tests that can dramatically shorten the required testing protocols (Nelson, 1990). And even if the acceleration is accidental, including additional covariate effects in the reliability analysis can often make more consistent estimates. Forms of ANOVA and regression exist that allow one to fit reliability models incorporating both censored data and covariate effects.

AN EXAMPLE SHOWING THE DIFFERENT APPROACHES

To demonstrate the application of these techniques to a measurement system, consider the case of the digital micrometer that was presented earlier. Assume that there are three failure modes, two electrical and one mechanical. The physical characteristics of these failures have been defined well, so it is easy to verify them when they occur. Assume that a study is conducted in which 100 measurement systems are checked every day for their performance in terms of repeatability and reproducibility. That is, each system is used in normal production operation by any of the pool of process operators who use it to adjust and set the manufacturing process. Once in a while the system fails due to one of the three modes previously mentioned. When this happens, the time is recorded and the device is removed from service. Table 13.1 shows the data from this record of events. Each entry is a failure time associated with a system and the mode by which it failed.

Table 13.1 Listing of uncensored failure times and modes.

Obs	i	Days	Type		Obs	i	Days	Type
1	1	24.171	elect_B		6	6	51.286	elect_B
2	2	42.137	elect_B		7	7	27.855	elect_B
3	3	13.134	elect_B		8	8	33.069	elect_B
4	4	18.424	elect_A		9	9	99.975	elect_B
5	5	21.478	elect_A		10	10	12.119	mech

(continued)

Table 13.1 Listing of uncensored failure times and modes.

Obs	i	Days	Type	Obs	i	Days	Type
11	11	6.508	elect_A	44	44	16.949	elect_A
12	12	12.682	mech	45	45	9.581	mech
13	13	10.430	mech	46	46	9.468	mech
14	14	2.726	elect_B	47	47	46.458	elect_B
15	15	8.107	mech	48	48	11.936	mech
16	16	3.998	elect_B	49	49	79.641	elect_B
17	17	6.283	mech	50	50	33.520	elect_B
18	18	8.777	elect_A	51	51	12.316	mech
19	19	155.140	elect_B	52	52	15.396	elect_A
20	20	12.072	mech	53	53	114.803	elect_B
21	21	41.435	elect_B	54	54	22.762	elect_A
22	22	28.701	elect_A	55	55	6.631	mech
23	23	22.208	elect_B	56	56	9.562	mech
24	24	4.771	mech	57	57	78.326	elect_B
25	25	2.226	elect_B	58	58	49.572	elect_B
26	26	44.491	elect_B	59	59	27.070	elect_A
27	27	5.585	elect_B	60	60	31.049	elect_A
28	28	7.250	mech	61	61	20.739	elect_B
29	29	21.278	elect_A	62	62	12.963	elect_A
30	30	44.803	elect_B	63	63	7.781	mech
31	31	7.028	elect_A	64	64	13.071	mech
32	32	17.867	elect_B	65	65	21.954	elect_A
33	33	24.585	elect_A	66	66	18.472	elect_B
34	34	39.101	elect_B	67	67	10.736	mech
35	35	14.923	elect_A	68	68	20.366	elect_A
36	36	197.279	elect_B	69	69	7.463	mech
37	37	7.811	mech	70	70	58.938	elect_B
38	38	7.488	elect_B	71	71	12.125	mech
39	39	54.453	elect_B	72	72	38.499	elect_B
40	40	5.904	mech	73	73	4.891	mech
41	41	35.815	elect_A	74	74	23.101	elect_A
42	42	62.225	elect_B	75	75	30.026	elect_A
43	43	5.105	mech	76	76	8.297	mech

(continued)

(continued)

Table 13.1 Listing of uncensored failure times and modes.

Obs	i	Days	Type	Obs	i	Days	Type
77	77	10.295	mech	89	89	30.511	elect_B
78	78	7.963	mech	90	90	11.414	elect_A
79	79	11.899	mech	91	91	12.912	mech
80	80	10.456	mech	92	92	7.991	elect_A
81	81	4.446	mech	93	93	27.360	elect_A
82	82	14.462	elect_B	94	94	12.514	elect_A
83	83	38.631	elect_B	95	95	11.988	elect_B
84	84	6.533	mech	96	96	7.908	elect_A
85	85	104.646	elect_B	97	97	9.181	mech
86	86	23.198	elect_A	98	98	24.530	elect_A
87	87	22.899	elect_B	99	99	16.311	elect_A
88	88	7.028	mech	100	100	26.673	elect_B

Assume that a parametric analysis using a two-parameter Weibull distribution is appropriate for each failure mode. If one fits the observed failure time to data sorted by failure mode, this will leave 30 values for mode 1, 36 values for mode 2, and 44 values for mode 3. Table 13.2 shows the details of each separate fit. Figures 13.2–13.4 show the fitted distribution for each of the three modes fitted separately.

Table 13.2 Results of the Weibull fits for the three modes.

Failure Mode	Weibull Scale	Weibull Shape
Electrical A	21.75	2.70
Electrical B	48.41	1.18
Mechanical	9.97	3.94

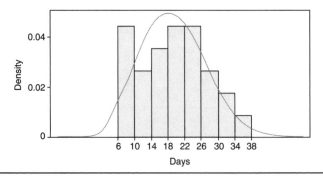

Figure 13.2 Failure times coded as elect_A.

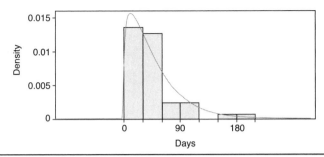

Figure 13.3 Failure times coded as elect_B.

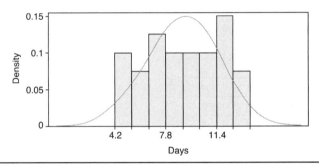

Figure 13.4 Failure times coded as mech.

It is also possible to fit these distributions even if some of the data are censored. Censored observations are removed from the test before their actual failure time is observed. Table 13.3 shows the data with some observations censored. Tables 13.4–13.6 and figures 13.5–13.7 show the new fits using this censored data for each of the three failure modes.

Table 13.3 Listing of failure and censored times.

i	Days	Type	Cens		i	Days	Type	Cens
1	24.171	elect_B	0		7	16.265	elect_A	0
2	26.243	elect_B	0		8	19.162	elect_A	0
3	11.204	mech	0		9	136.751	elect_B	0
4	6.336	elect_B	0		10	26.708	elect_B	0
5	27.787	elect_B	0		11	10.430	mech	1
6	27.855	elect_B	0		12	25.345	elect_B	0

(continued)

Table 13.3 Listing of failure and censored times.

i	Days	Type	Cens		i	Days	Type	Cens
13	13.852	elect_A	0		36	5.105	mech	0
14	9.412	elect_A	0		37	15.125	elect_B	1
15	27.288	elect_A	0		38	15.620	mech	0
16	155.140	elect_B	0		39	83.623	elect_B	0
17	2.542	mech	0		40	9.284	mech	0
18	16.144	elect_A	0		41	79.641	elect_B	0
19	2.325	elect_B	0		42	50.856	elect_B	0
20	23.709	elect_B	0		43	28.799	elect_A	0
21	2.226	elect_B	0		44	14.479	elect_B	0
22	18.866	elect_A	1		45	64.761	elect_B	0
23	11.207	mech	0		46	6.631	mech	0
24	10.127	elect_A	0		47	42.739	elect_B	0
25	4.407	elect_B	0		48	53.187	elect_B	0
26	7.028	elect_A	0		49	17.533	elect_A	0
27	85.424	elect_B	0		50	120.508	elect_B	0
28	26.296	elect_B	0		51	20.739	elect_B	0
29	7.773	elect_A	0		52	28.295	elect_A	1
30	34.439	elect_B	0		53	12.893	elect_A	0
31	7.811	mech	1		54	17.626	elect_A	1
32	7.740	elect_A	0		55	9.150	elect_A	0
33	20.022	elect_A	1		56	10.736	mech	0
34	114.437	elect_B	0		57	7.981	elect_A	0
35	122.369	elect_B	0		58	2.149	elect_B	1

(continued)

(continued)

Table 13.3   Listing of failure and censored times.

i	Days	Type	Cens		i	Days	Type	Cens
59	6.263	mech	0		80	7.817	elect_B	0
60	19.151	elect_B	0		81	9.181	mech	0
61	4.891	mech	0		82	79.990	elect_B	0
62	5.459	mech	0		83	145.978	elect_B	0
63	136.007	elect_B	0		84	33.448	elect_A	0
64	95.068	elect_B	0		85	15.509	elect_B	0
65	12.976	elect_B	0		86	324.607	elect_B	1
66	11.899	mech	0		87	16.254	elect_A	1
67	27.101	elect_A	0		88	10.069	mech	1
68	4.841	elect_A	1		89	14.982	elect_A	0
69	16.508	elect_A	0		90	108.011	elect_B	1
70	10.512	mech	0		91	20.447	elect_B	1
71	104.646	elect_B	0		92	10.048	mech	1
72	21.995	elect_B	0		93	13.816	mech	1
73	19.501	elect_B	0		94	150.256	elect_B	0
74	9.128	mech	0		95	11.710	mech	0
75	16.285	elect_B	1		96	30.546	elect_A	0
76	12.912	mech	1		97	26.121	elect_A	1
77	5.208	mech	0		98	25.770	elect_A	0
78	16.929	elect_A	0		99	10.283	elect_A	1
79	23.564	elect_A	0		100	11.479	elect_A	1

Table 13.4 Details of the fit for failure mode elect_A.

Summary of Fit	
Observations Used	32
Uncensored Values	23
Right Censored Values	9
Maximum Loglikelihood	−23.11587

Weibull Parameter Estimates				
			Asymptotic Normal 95% Confidence Limits	
Parameter	**Estimate**	**Standard Error**	**Lower**	**Upper**
EV Location	3.1116	0.0858	2.9434	3.2798
EV Scale	0.4101	0.0674	0.2972	0.5660
Weibull Scale	22.4568	1.9270	18.9806	26.5698
Weibull Shape	2.4381	0.4007	1.7667	3.3647

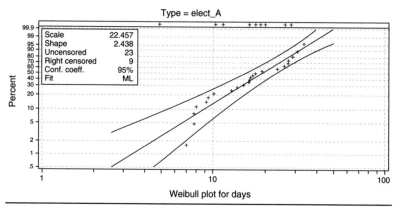

Figure 13.5 Graphical representation of the fit for failure mode elect_A.

Table 13.5 Details of the fit for failure mode elect_B.

Summary of Fit	
Observations Used	45
Uncensored Values	39
Right Censored Values	6
Maximum Loglikelihood	−65.70469

Weibull Parameter Estimates				
			Asymptotic Normal 95% Confidence Limits	
Parameter	**Estimate**	**Standard Error**	**Lower**	**Upper**
EV Location	4.2386	0.1636	3.9179	4.5592
EV Scale	0.9939	0.1226	0.7805	1.2657
Weibull Scale	63.3096	11.3392	50.2962	95.5105
Weibull Shape	1.0061	0.1241	0.7901	1.2812

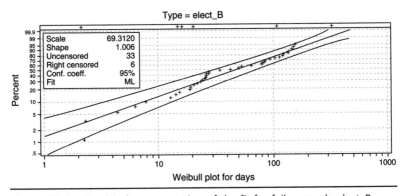

Figure 13.6 Graphical representation of the fit for failure mode elect_B.

Table 13.6 Details of the fit for failure mode mech.

Summary of Fit	
Observations Used	23
Uncensored Values	17
Right Censored Values	6
Maximum Loglikelihood	−16.30938

Weibull Parameter Estimates				
			Asymptotic Normal 95% Confidence Limits	
Parameter	**Estimate**	**Standard Error**	**Lower**	**Upper**
EV Location	2.4194	0.0870	2.2489	2.5899
EV Scale	0.3586	0.0724	0.2414	0.5328
Weibull Scale	11.2393	0.9777	9.4775	13.3285
Weibull Shape	2.7884	0.5631	1.8770	4.1423

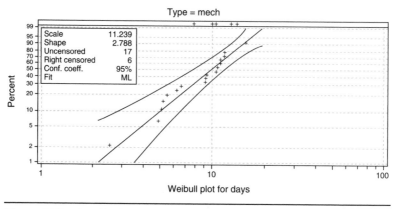

Figure 13.7 Graphical representation of the fit for failure mode mech.

The technician who is analyzing this data realizes that she may have an additional source of data to add to the original analysis. There were actually 500 devices in use, but she had purposefully selected only a specific subset of 100 systems because the others were used with higher voltage power supplies. Higher voltage is known to increase the failure rate of the micrometers for electronic failure mode number three. Table 13.7 shows the data from the failures for each of the five voltage levels. Table 13.8 shows an analysis in which the scale of the Weibull distribution is allowed to vary in a linear fashion with voltage (Allison, 1995).

Table 13.7 Variation of failure times with voltage.

Obs	Voltage *i*	Device	Days	Cens
1	50	1	65.2306	0
2	50	2	74.1809	0
3	50	3	58.6578	0
4	50	4	55.0808	0
5	50	5	98.8089	0
6	50	6	66.4864	0
7	50	7	68.2175	1
8	50	8	73.2116	0
9	50	9	99.5091	0
10	50	10	91.2246	1
11	60	1	53.6950	0
12	60	2	43.1658	0
13	60	3	21.8440	0
14	60	4	50.6390	0
15	60	5	55.5055	0
16	60	6	33.8621	0
17	60	7	25.1787	0
18	60	8	44.8578	0
19	60	9	40.6626	1
20	60	10	16.6224	0
21	70	1	20.3152	0
22	70	2	41.0606	0
23	70	3	37.2173	0
24	70	4	30.3730	0
25	70	5	63.4762	0
26	70	6	13.9461	0
27	70	7	32.5912	0
28	70	8	14.7492	0
29	70	9	27.8498	0
30	70	10	32.8585	0
31	80	1	19.2798	1
32	80	2	40.3668	0

(continued)

(continued)

Table 13.7 Variation of failure times with voltage.

Obs	Voltage *i*	Device	Days	Cens
33	80	3	47.8893	1
34	80	4	8.7565	0
35	80	5	30.3349	0
36	80	6	46.7175	0
37	80	7	25.9721	0
38	80	8	52.2624	0
39	80	9	10.6040	0
40	80	10	17.6480	0
41	90	1	43.8002	1
42	90	2	14.7795	0
43	90	3	45.4800	1
44	90	4	4.7052	0
45	90	5	20.5575	0
46	90	6	24.8841	0
47	90	7	22.3749	0
48	90	8	24.5662	0
49	90	9	44.9666	0
50	90	10	3.0329	0
51	100	1	38.5661	0
52	100	2	24.2161	0
53	100	3	32.3043	0
54	100	4	10.2782	0
55	100	5	31.8690	1
56	100	6	38.8052	0
57	100	7	34.2254	0
58	100	8	22.1315	0
59	100	9	28.5306	1
60	100	10	28.1082	0

Table 13.8   Details of the fit for the voltage-dependent failures.

Parameter	DF	Estimate	Standard Error	Lower Limit	Upper Limit	ChiSq	Pr > ChiSq
Intercept	1	5.0148	0.2490	4.5268	5.5028	405.67	<.0001
Voltage	1	–0.0167	0.0032	–0.0230	–0.0103	26.64	<.0001
Scale	1	0.4487	0.0525	0.3568	0.5644	—	—
Weibull Shape	1	2.2284	0.2608	1.7717	2.8029	—	—

CLOSING REMARKS

Because measurement systems are composed of multiple components, their performance can be every bit as complex as the manufacturing or service processes to which they are applied. Because of this complexity, many of the tools and approaches that have been shown to be useful in mainstream quality applications are just as valuable for MSA studies. Such tools include process diagrams, fishbone diagrams, capability studies, control charts, designed experiments, and brainstorming (Berk, 1993).

Perhaps the subtlest difference between production process and measurement process management is that measurement impacts are usually indirect. It is the information they provide in order to enable decisions that is of utmost importance. Unless one knows the details of the performance of a measurement system, one cannot make good choices about the deployment, maintenance, and improvement of the measurement system. The MSA is the primary way in which one gains the information that is needed to understand these performance characteristics. It should be assumed that all measurement systems are fallible and have error.

Chapter 13 Take-Home Pay

1. Measurement systems can fail completely or through sufficient deterioration.

2. The reliability of measurement systems can be studied and improved.

3. Typically, failure modes must be studied separately.

4. Data for the measurement system's reliability can be censored or uncensored.

5. Analysis can also allow the reliability to be a function of underlying variables.

14

A Review of Progress with an Eye to the Future

CURRENT STATUS OF MSA

Measurement Impacts

Measurements are the only way in which one can interact with the processes and systems that are important to manufacturing and business. Whether these interactions are for the purpose of gaining information or for inducing control, measured values provide the linkage between our concepts and reality. The quality of this information is directly related to the potential errors that can occur in the measurement process. The quality of decisions based on the information that is provided by measurements is thus also directly related to these errors.

The classification of products as to their conformance to specifications can easily be affected by measurement error. A measurement error that overestimates the true value of a product can lead to false scrap and mistaken reworking. On another product the measurement error may underestimate the true value of the product and prevent the detection of a serious nonconformance. Because these kinds of errors can occur with every measurement that is taken, even small error sizes can lead to large economic losses. Product values near the tolerance limits are more sensitive to measurement effects than others, so it is a good strategy to have high-capability processes to mitigate the impact of measurement errors.

The same kinds of measurement errors can also affect other activities. Process control tries to maintain the process as closely as possible to a target performance. Corrective actions can be sporadic and relatively large, as is practiced in a statistical process; assurance of it can be continuous and relatively small, as in feedback control. In either case, these corrective actions must be based on measurements that can be in error. Thus the effectiveness of control can be reduced and even made negligible. Depending on the control algorithms that are employed, the measurement errors can be a small nuisance or a major disruption, but they always add to the cost.

Laboratory experiments and in-shop studies are also affected by measurement errors. This may simply add to the time that one has to take to arrive at correct conclusions, or it may actually make the journey from ignorance to knowledge nearly impossible. In manufacturing and service industries, where time is often of the essence, large measurement errors disrupt the entire learning process and cause many expensive delays. In complex situations in which many causes interact in a subtle fashion, the introduction of measurement error into the process can lead to innumerable problems. In many sociological and political processes the measurements can be so poor as to make learning all but impossible except in a long-term, historical view.

The Standard Measurement Study

Because measurement errors can cause so many issues and negatively affect costs, it is prudent to study the performance of measurement systems in order to manage their impact. In many industries an uncertainty analysis is recommended. In an uncertainty analysis approach, one makes a list of all possible causes of measurement error for a particular measurement system. Such a list will usually include personnel effects, method effects, material effects, and environmental effects. The next step in the uncertainty analysis is to estimate the contribution of each component on the list either by data collection or by engineering reasoning. Often these component contributions are combined into one overarching total uncertainty that summarizes the performance of a measurement system with one number.

Because there are too many measurement systems in use to perform a complete uncertainty analysis on each one, a shortened approach called a measurement systems analysis (MSA) is often performed. Such an analysis concentrates on the estimation of three sources of measurement errors that are often found to be dominant in real systems. A full factorial of all combinations of 10 parts measured by three appraisers for three repetitions is recommended for the study. These parts are assumed to represent the full range of part variability that can be expressed in the process, and measurements are taken independently of one another in the study.

An MSA often employs the technique of analysis of variance to provide the estimates of the three components of variation that are captured in the study: parts, appraiser, and repeatability. The appraiser variability is also called the reproducibility. Similar to uncertainty analysis, the two components of measurement error are combined into one summary measurement called gage RandR. All estimates are given as standard deviations, multiples of standard deviations (usually six), or percentages of tolerance ranges.

Often measurement systems are judged adequate if the percentage tolerance is less than 10 percent, marginally acceptable if between 10 and 30 percent, and unacceptable if over 30 percent. Even though these two components of uncertainty may not capture all the measurement system error, it is still valuable to study RandR for a wide variety of systems and applications.

Problems and Extensions of the Standard MSA

The standard MSA is valuable and valid for many measurement systems, but there are exceptions that can be important. The estimate of appraiser variability that is obtained from the standard MSA can be quite unreliable and can have large uncertainties associated with it. One must take this variability into account in order to make proper decisions on measurement system performance and improvement. One way to improve this estimate is to increase the number of appraisers involved in the study. Increasing the number of repetitions and the number of parts can also affect the estimate of reproducibility but not nearly to the same degree of impact.

Fortunately, the ANOVA components analysis approach can be modified in many ways. For example, it is possible to add the analysis of interactions to the ANOVA without any increase in modification of the basic data collection scheme. And yet the interaction terms can add a great deal to the understanding of the measurement system performance. The removal of interaction effects can also directly reduce the estimate of repeatability because of its reduction of the residual error that is the basis of the estimation procedure. And any decrease in the repeatability directly makes the uncertainty less in the estimates of part and appraiser variance components. In general, the incorporation of more uncertainty sources into the ANOVA study helps give a more complete understanding of the measurement error and gives more precise estimates of all components.

Deformative, Destructive, and Dynamic Measurement Studies

The ANOVA approach can be modified so that it can deal effectively with the very difficult situations in which the true part value does not repeat itself from trial to trial. In some cases this part may deform with measurement, as when an ammeter draws current during operation. In other cases the part or its pertinent characteristic may be completely destroyed, as when a sample is burned in a calorimeter. In still other cases the part is not affected, but it is difficult or impossible to present the same part to the measurement system in the same way twice. An in-line weight scale may operate only in a dynamic mode on a real product that is processed and made inaccessible as soon as it is weighted.

The part change effect and the measurement effects may be separated by clever experimental designs or through the use of statistical models. The ANOVA can be extended to handle continuous effects like time and temperature, which can estimate and isolate the part change effect. Some patterns that can be easily separated include linear effects, piecewise linear effects, and sinusoidal effects. The patterns can be generated on the basis of engineering knowledge of the way in which the parts change or from statistical searches for patterns. With a little care so as not to go too far in the fitting process, this approach can give much better estimates of the underlying measurement system performance.

Design of MSAs

The standard MSA is based on a very simple full factorial design in parts, appraisers, and repetitions. When this is applied, it turns out that most of the trials go to estimating the repeatability component and very little (relatively speaking) go to the estimation of part and appraiser effects. With this understanding, it is possible to decrease the number of runs in at least two ways. In either case, the purpose is to provide separable estimates of all required components with enough elements for each source to provide reliable estimates of these same components.

One way to reduce the number of runs is to simply reduce the number of repetitions. This can be done consistently across all combinations to cut the size to one-third of the standard size experiment. Or it can be done on a few randomly selected combinations to provide a reduction up to the full two-thirds. Another way to reduce the runs is to employ less than full factorials. Fractional factorials, orthogonal arrays, and computer-generated optimal designs can all be used to successfully reduce the array sizes. For example, a study with six sources, each with two levels, might require 64 runs, whereas a smaller array of only eight runs may provide similar estimates.

Attribute Measurement Systems Analyses

Attribute measurement systems are quite common and provide simple, discrete, often yes-no results. This loss of resolution can make it difficult to do a good study. The standard attribute MSA takes a simplistic approach in which 50 parts of various true part values are presented to the systems with three appraisers and three repetitions. There is waste in this approach if the parts are not selected carefully. Also, the targets for attribute measurement systems performance are much weaker than those for variables measurement systems. If one wishes to extend the method to more uncertainty sources, a different method is recommended that generalizes the analytic method.

Two approaches are given that are oriented toward improving attribute measurement systems. One approach is based on maximum likelihood analysis of simple repeatability and reproducibility. Another approach is based on logistic regression and allows one to drastically extend the analysis to many more factors. The logistic regression approach also allows one to include continuous effects in the analysis. Mimicking the approach given for variable systems, it is possible to use this logistic regression approach to extend attribute analyses to deformative, destructive, and dynamic situations. In this way, one can get just as much power and clarity for attribute measurement studies as for variable studies.

Measurement Performance and Reliability

MSAs are often done merely to satisfy an audit or a customer requirement. A better approach is to evaluate the costs of poor measurement performance and use this to track and motivate additional work in this area. Generally, for any application of a measurement system there are errors stemming from false-positive and false-negative results. By establishing costs for these two kinds of errors, it is possible to compute the expected loss of risk due to poor measurement systems performance. Applied consistently and rationally, this computation of losses can be used as the basis for a measurement management program.

The physical nature of measurement systems means that they can also be subject to failure. This failure can be complete, as when the equipment fails to operate, or it can be a failure due to a sufficient deterioration in performance. Analysis can be done in either case to provide estimates of the reliability of the measurement. Methods are available for data consisting of all complete failures, for mixtures of failure and censored data, and for accelerated life tests in which the failure rate can depend on background factors such as temperature. Because reliability information on measurement systems can be even rarer than data on measurement systems performance, it is critical to use the data as efficiently as possible.

FUTURE DIRECTIONS FOR MSAS

Clearly, if the methods presented in this book are applied, there could be a revolution in measurement systems evaluation and improvement. Specifically, there are at least five directions in which the management of measurement uncertainty is likely to evolve. These directions include (1) the practice of more simple measurement studies, (2) the practice of more complete uncertainty analyses, (3) the analyses of complexes of measurement systems, (4) the practice of MSA to nonmanufacturing applications, and (5) the improved design of measurement systems.

In most industrial and manufacturing settings it is difficult to run enough MSAs of even the simplest design. Oftentimes compromises are made in the quality of the studies, and all too often, follow-up studies are not done. Part of the reason for this is the expense of collecting 90 measurements with three appraisers. By using the methods of MSA design to reduce the runs, it is quite possible to get equally adequate information from studies that are less than one-quarter of the standard size. And if the precision of the appraiser estimate is not considered adequate with three appraisers, they could be omitted. In this way, it should be possible to run more MSA studies with the same budgets. If the impact of poor measurement performance is tracked, the budget should also increase as well.

Simple measurement studies provide useful insight into the performance of the systems that can be used to improve them. But they fall far short of the full uncertainty analysis that really characterizes the impact of the measurement error. With the application of fractional factorial designs for uncertainty studies, it should be possible to get a more complete understanding of the measurement effects without dramatically increasing the size or cost of the data collection. The relationship between the RandR results and the full uncertainty depends on the particular applications, but for many applications it can be correctly assumed that the RandR is merely a minor fraction of the full amount. So a lot of improvement is almost sure to result from the running of more uncertainty analysis studies.

A single measurement system is often complicated in itself, but in most applications it does not work alone. It is common to have a sequence of measurements that determine the quality chain of a product or service. For example, there might be a linear measurement of area, a weight measurement, and an attribute aspect measurement that are applied as part of the inspection management program in one particular process in an industry. Each measurement process can be analyzed separately by the methods in this book, but it might be more direct to study the combined effects of all three systems. This viewpoint is one that treats the situation as a complex of measurement systems. In some situations it can be more efficient to analyze the whole complex to determine its capability. In other cases the same measurement system might be used several times in order to make the system redundant. Treatment of complexes of systems could be very important in the development of better guarantees for complicated situations.

Arguably, the most important future direction of MSA is to include nonmanufacturing systems in the scope of the studies. Measurements are made in every system, whether it is an accounting audit, a sales campaign, or the provision of access to an Internet provider. There is no reason why the approaches described in this book cannot be performed on these nonmanufacturing systems. It is reasonable to expect that the errors of these systems will be relatively large because they have been considered inaccessible to study for so long. Without good studies it is very difficult to motivate and direct measurement improvement efforts.

Finally, the more complete and better understanding of the full measurement system uncertainty should be valuable in another direction as well. Once the performance of the active systems has been determined and characterized, it is natural to use this information to improve the design of the system. Sometimes it is not feasible to modify the current system, but regardless, the information should be useful in designing replacement and new systems of measurement. Knowledge is almost always the key to saving money and making improvements. Assuming that most measurement systems will have a measurement uncertainty that is larger than desired, it is of great advantage to design the uncertainty out of the measurement system in the first place.

Regardless of the velocity that is shown in each of these directions, it seems quite clear that the analysis of measurement systems will become more and more important in achieving a good bottom-line result. As the impact of these errors is better understood, it will become too costly to sustain large measurement uncertainty. And as more systems are embedded in the manufacturing and service processes, it will be necessary to use many of the techniques described in this book to reduce the measurement errors. The future of MSA appears quite likely to heat up substantially in the near future for these reasons and others that are not seen with as much clarity from the current vantage point.

Chapter 14 Take-Home Pay

1. Measurement error has potential impact on all decisions.

2. Standard MSA studies are good but can be improved.

3. Multiple error sources enable more complete understanding.

4. Multiple error sources yield more precise estimates of all quantities.

5. There are methods for destructive, deformative, and dynamic MSA.

6. There are much better methods for attribute MSA.

7. The design of MSA studies is the key to valid and efficient studies.

8. Measurement value can be captured and can justify improvements.

9. Measurement systems must be kept reliable.

10. In the future, measurement management will become ever more critical.

Bibliography

AIAG (Automobile Industry Action Group). (2002). *Measurement Systems Analysis*. Third Edition. DaimlerChrysler, Ford Motor Company and General Motors Corporation.

Allison, Paul. (1995). *Survival Analysis Using SAS*. Cary, NC: SAS Publishing.

Bergeret, F., S. Maubert, P. Sourd, and F. Puel. (2001). "Improving and Applying Destructive Gauge Capability" in *Quality Engineering* V14 No. 1, pp. 59–66.

Berk, Joseph, and Susan Berk. (1993). *Total Quality Management*. New York: Sterling Publishing.

Box, George E. P., David E. Coleman, and Robert V. Baxley Jr. (1997). "A Comparison of Statistical Process Control and Engineering Process Control" in *Journal of Quality Technology* V29 No. 2, pp. 128–130.

Boyles, R. A. (2001). "Gauge Capability for Pass-Fail Inspection" in *Technometrics* V43 No. 2, pp. 223–229.

Burdick, Richard K., Connie M. Borror, and Douglas C. Montgomery. (2003). "A Review of Methods for Measurement Systems Capability Analysis" in *Journal of Quality Technology* V35 No. 4, pp. 342–354.

De Mast, Jeroen, and Albert Trip. (2005). "Gauge R&R Studies for Destructive Measurements" in *Journal of Quality Technology* V37 No. 1, pp. 40–49.

D'Errico, John R., and Nicholas A. Zaino Jr. (1988). "Statistical Tolerancing Using a Modification of Taguchi's Method" in *Technometrics* V30 No. 4, pp. 397–405.

Ding, Jie, Betsy S. Greenberg, and Hirofumi Matsuo. (1998). "Repetitive Testing Strategies When the Testing Process Is Imperfect" in *Management Science* V44 No. 10, pp. 1367–1378.

Dixon, Wilfred J., and Frank J. Massey Jr. (1983). *Introduction to Statistical Analysis*. New York: McGraw-Hill.

Doganaksoy, Necip, and Gerald J. Hahn. (1996). "Evaluating the Potential Impact of Blending on Product Consistency" in *Journal of Quality Technology* V28 No. 1, pp. 51–60.

George, Michael. (2002). *Lean Six Sigma: Combining Six Sigma Quality with Lean Speed*. New York: McGraw-Hill.

Harrington, H. J. (1991). *Business Process Improvement*. New York: McGraw-Hill.

Henley, Ernest J., and Hiromitsu Kumanoto. (1981). *Reliability Engineering and Risk Assessment.* Englewood Cliffs, NJ: Prentice Hall.

Hillier, Frederick S., and Gerald J. Lieberman. (1974). *Operations Research.* San Francisco: Holden-Day.

Lawless, J. F., and C. Nadeau. (1995). "Some Simple Robust Methods for the Analysis of Recurrent Events" in *Technometrics* V37 No. 2, pp. 158–168.

Lindlay, D. V. (1985). *Making Decisions.* Second Edition. New York: John Wiley & Sons.

Little, Ramon C., George A. Millilken, Walter W. Stroup, and Russell D. Wolfinger. (1996). *SAS System for Mixed Models.* Cary, NC: SAS Institute.

Montgomery, Douglas C. (2000). *Design and Analysis of Experiments.* Fifth Edition. New York: John Wiley & Sons.

Myers, Raymond H., and Douglas C. Montgomery. (2002). *Response Surface Methodology.* Second Edition. New York: John Wiley & Sons.

Myers, Raymond H., Douglas C. Montgomery, and G. Geoffrey Vining. (2001). *Generalized Linear Models.* New York: John Wiley & Sons.

Nelson, Wayne. (1982). *Applied Life Data Analysis.* New York: John Wiley & Sons.

———. (1990). *Accelerated Testing.* New York: John Wiley & Sons.

Ott, Lyman. (1984). *An Introduction to Statistical Methods and Data Analysis.* Second Edition. Boston: PWS Publishers.

Phillips, Aaron R., Rella Jeffries, Jan Schneider, and Stanley P. Frankoski. (1997). "Using Repeatability and Reproducibility Studies to Evaluate a Destructive Test Method" in *Quality Engineering* V10 No. 2, pp. 283–290.

Raktoe, B. L., and W. T. Federer. (1981). *Factorial Designs.* New York: John Wiley & Sons.

Rubenstein, R. Y. (1981). *Simulation and the Monte Carlo Method.* New York: John Wiley & Sons.

Runger, George C., and Joseph J. Pignatiello Jr. (1991). "Adaptive Sampling for Process Control" in *Journal of Quality Technology* V23 No. 2, pp. 135–155.

Ryan, Thomas P. (1989). *Statistical Methods for Quality Improvement.* New York: John Wiley & Sons.

Schilling, E. G. (1982). *Acceptance Sampling in Quality Control.* New York: Marcel Dekker.

Schwartz, Mischa, and Leonard Shaw. (1975). *Signal Processing.* New York: McGraw-Hill.

Searle, S. R. (1971). *Linear Models.* New York: John Wiley & Sons.

Snedecor, George W., and William G. Cochran. (1967). *Statistical Methods.* Sixth Edition. Ames: Iowa State University Press.

Stamatis, D. H. (2003). *Failure Mode and Effects Analysis: FMEA from Theory to Execution.* Second Edition. Milwaukee, WI: ASQ Quality Press.

Index

parts, 17, 28
 changing number of, 35–37
 reference, effects of poorly
  chosen, 122–126
pilot programs, 174
prediction probabilities, for attribute
  MSA, 133–134
process adjustment, 6–8
 vs. process control, 8
process control
 effects, 8–10
 vs. process adjustment, 8
product change patterns, finding,
  76–78
product classification
 defined, 2–3
 errors, 2–6

R

randomization, complete, 86
RandR, 20. *See also* repeatability;
  reproducibility
 application of logistic regression
  to, 148–156
 computation of, 19
 variability of, 40–41
reference parts, effects of poorly
  chosen, 122–126
reliability
 defined, 177
 example showing differing
  approaches to, 180–191
 failure-based, for measurement
  systems, 177–180
 measuring, 197
repeatability, 16–17. *See also* RandR
 estimate of standard deviation of,
  18

evaluating variability in estimates
  of, 23–27
forcing unknown effects into,
  108–112
in full factorial approach, 99–101
repeats
 changing number of, 38–40
 effects of nonindependence in, 121
repetitions, 17
reproducibility, 16–17. *See also*
  RandR
 for dynamic measurement
  systems, 83–84
 standard deviation of, 19
residuals, 74
risk. *See also* measurement risk
  management programs
 assessment, 170–171
 cost component of, 167–169
 defined, 165
 evaluation of, attribute gages and,
  169–170
 on product sorting, 165–167
risk improvement, approaches for,
  173–174

S

sample size, attaining adequate,
  105–108
sampling plans, 173
Shewhart process control charts,
  construction of, 8–9
single repeat, using, for MSA studies,
  103–104
Six Sigma, 5
standard deviation
 of appraiser effect, 19
 of repeatability, 18